Taunton's

WIRING COMPLETE

EXPERT ADVICE FROM START TO FINISH

Text by Michael W. Litchfield and Michael McAlister, © 2008 by
The Taunton Press, Inc.
Photographs © 2008 by The Taunton Press, Inc., except as noted.
Illustrations © 2008 by The Taunton Press, Inc.

The Taunton Press, Inc.,
63 South Main Street, PO Box 5506,
Newtown, CT 06470-5506
e-mail: tp@taunton.com

Editors: Matthew Teague, Helen Albert
Copy editor: Candace B. Levy
Indexer: Jay Kreider
Jacket/Cover design: Kimberly Adis
Interior design: Kimberly Adis
Layout: Cathy Cassidy
Illustrator: Trevor Johnston
Photographer: Michael Litchfield, except where noted.

Library of Congress Cataloging-in-Publication Data

Litchfield, Michael.
 Wiring complete : expert advice from start to finish / Michael Litchfield
 and Michael McAlister.
    p. cm.
 Includes bibliographical references and index.
 ISBN 978-1-56158-815-2 (alk. paper)
 1. Electric wiring, Interior. 2. Dwellings--Electric equipment.
   I. McAlister, Michael. II. Title.

TK3285.L54 2008
621.319'24--dc22
                2007048861

Printed in the United States of America
10 9 8 7 6 5 4 3 2 1

Dozens of people made this book possible. We are especially indebted to the electricians and builders who got us onto job sites and allowed us to photograph them at work. Thanks to old friend Rafael Maldonado of Blue Electric in Berkeley, California, for sharing his vast knowledge and his crew, particularly Isaac Castro, Jorge Dominguez, and Carlos Guerrero. Huge thanks to the unflappable Daniel Kealey, Nathan Parker, and Din Abdullah of MRM Electrical, Berkeley. Likewise, Kevin McCarthy, Simon Jordan, Ki Soo An, and Jimmy Stuart. A million thanks to Mike Zelinka and colleagues at Laner Electric in Richmond, California.

We are grateful to Jana Olson, Karen Cornell, and John Nicoles of Omega Too Lighting in Berkeley for the excellent sequences of lamp and chandelier repair. Thanks also to Nowell's Antique Lighting in Sausalito, California.

Hats off to Jamie Carlen and Chris D'Andrea of Jamie Carlen & Co., Berkeley; and Chip Harley, Jesus Beltran, and Gelber Guzman of Holland-Harley. Thanks to supremely talented architect Gary Earl Parsons of Berkeley and his associate Cecil Lee.

Thanks to Roger Robinson of the Star Inspection Group, Oakland, California, for letting us cherry-pick his photo archive of code violations; Casablanca Fans for product shots; Muffy Kibbey for photos of the handsome kitchen in the planning chapter; and Ken Gutmaker for photos that first appeared in *Renovation, 3rd Edition* (Taunton, 2005).

Thanks to homeowners Catherine Moncrieff and Stuart Brotman, Martha and Dean Rutherford, Judith and Stanley Lubman, Laurie Case, and Andy Baker.

Finally, high-fives to the Taunton Press family and friends, including Executive Editor Helen Albert, Wendi Mijal, Matthew Teague, Jennifer Peters, Nicole Palmer, Julie Hamilton, Sandor Nagyszalanczy, Rex Cauldwell, Clifford Popejoy, Joseph Truini, Mike Guertin, and Fernando Pagés Ruiz.

—Michael Litchfield and Michael McAlister

Wiring is inherently dangerous. Using hand or power tools improperly or ignoring safety practices can lead to permanent injury or even death. Don't try to perform operations you learn about here (or elsewhere) unless you're certain they are safe for you. If something about an operation doesn't feel right, don't do it. Look for another way. We want you to enjoy working on your home, so please keep safety foremost in your mind.

contents

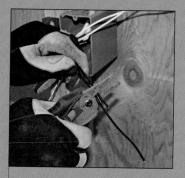

≫ ≫ ≫ ≫

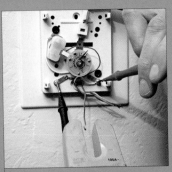

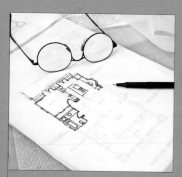

》 》 》 》

INTRODUCTION

Electrical wiring requires attention to detail, patience, and a little dexterity, but it's nothing the average homeowner can't tackle. Before you buckle on that tool belt, however, take a few moments to read the first section, which provides an overview of electrical systems and a handful of essential safety tips. Section Two walks you through the tools you'll need and various techniques you'll use again and again. The remainder of the book takes you through every step of the wiring process—from replacing an old light fixture to wiring an entire house.

Before you buy tools and materials, however, check with local building code authorities. Although most local building codes do not forbid an owner's doing his or her own electrical work, most require a rough inspection—that is, before wires are connected to switches, receptacles and so on—and a final inspection when everyting is wired, trimmed, and tested. Besides, building inspec-

tors are usually knowledgeable: They can tell you if local codes conform to the National Electrical Code® or, if not, how they vary. Finally, check with your insurance agent to make sure that doing your own electrical work won't jeopardize your homeowners insurance coverage.

WORKING WITH ELECTRICITY

BEFORE WORKING WITH electricity, you should have a basic understanding of how it works. This chapter is designed to give you a quick overview of the electrical system in your home, including the major components. Since grounding is essential to keeping you safe, we cover that in detail. We'll also show how to do a basic inspection of your home for wiring problems. Respecting the power of electricity is essential to working safely. Always follow the instructions carefully, use appropriate safety equipment, and when in doubt consult a licensed electrician.

Before beginning work, check with local building authorities to make sure local regulations allow you to do your own work and that you are conforming to local code requirements.

BASICS

CUTTING POWER

TESTING

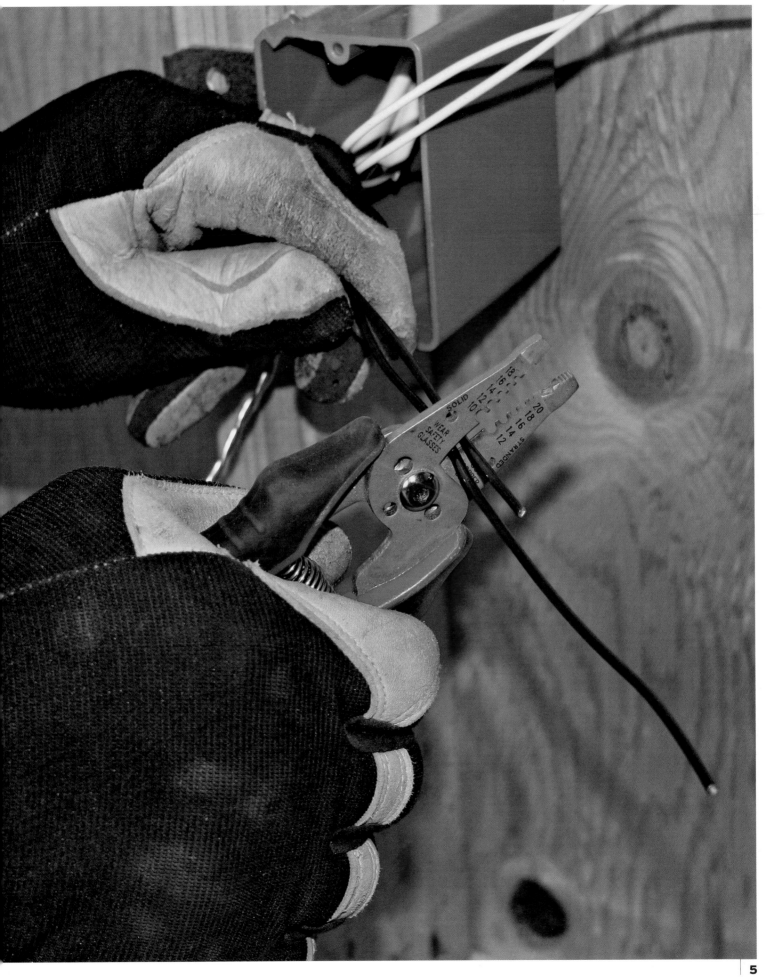

UNDERSTANDING ELECTRICITY

The easiest way to understand how electricity flows is to visualize water flowing through a pipe. Electricity (flowing electrons called *current*) moves through a circuit like water in a pipe. Just as the flow of water is measured in gallons per minute, the electrical flow of electrons is measured in "amperes" or amps. Water pressure is measured in pounds per square inch, and the force of electrons in a circuit is measured in *volts*. The larger the pipe, the more water that can flow through it; likewise, the larger wires allow a greater flow of electrical current. Too much water can rupture a pipe. Wiring that is too small will resist the flow of current. If that resistance (measured in *ohms*) is too great, the wires can melt and cause and start a fire.

AC (alternating current) electrical systems can be thought of as a loop that runs from the generation point (or power source) through utilization equipment (a light bulb for instance) and back to the generation point. Along the way it may cross the country through great transmission lines, through the power lines on your street, and through the cables in the walls of your house. In your home, the main loop, that is the service to your home, is broken into smaller loops called circuits. Typically, a *hot* wire (usually black or red) carries current to the utilization equipment from the service panel, and a *neutral* wire (typically white or light gray) carries current back to the service panel.

CIRCUIT BASICS

Electricity always flows in a circuit. The hot wire carries the current from the power source to the fixture and the neutral wire returns it to the power source.

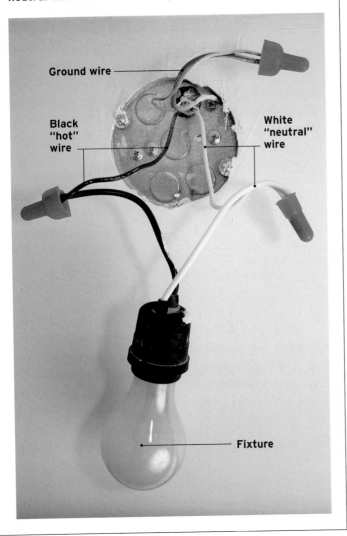

Ground wire

Black "hot" wire

White "neutral" wire

Fixture

Key Terms

Power (Volt-Amps or VA)
The potential in the system to create motion (motors), heat (heaters), light (fixtures or lamps), etc. Volt-amps = available volts x available amps (VA).

Watts
Are a measure of power consumed. Watts are very similar to volt-amps, the main difference being that some energy is lost through heat and/or inherent inefficiencies in equipment.

Voltage
Voltage is the *pressure* of the electrons in a system. Voltage is measured in *volts*.

Amperes (Amps)
The measure of the volume of electrons flowing through a system (current).

Current
The flow of electrons in a system. Current is measured in amperes (amps). There are two types of current: DC (direct current) and AC (alternating current). Typically AC is found in homes and buildings.

Ohms
Ohms is the measure of resistance to the flow of electrons (current) in a material (like a cable). The higher the resistance the lower the flow of electrons.

WORKING SAFELY

To work safely with electricity you must respect its power. If you understand its nature and heed the safety warnings in this book—especially shutting off the power and testing with a voltage tester to make sure power is off—you can work with it safely. The cardinal rule of home-improvement projects, which goes double for electrical work, is: Know your limitations. Unless you have previous experience doing electrical work and feel confident about your skills, you should leave certain projects to a pro. Working inside a service panel or removing its cover can be especially dangerous. There is an area around the main breaker that remains hot even after the breaker is set to the off position. Never attempt to remove or repair the main service panel or the service entrance head. When in doubt, call a licensed electrician rather than risk harm.

For most projects, careful attention to following the instructions in this book will keep you safe and ensure good results. Where there is a particular risk in a project, we will call attention to it with the following symbol:

WARNING
Please read this information carefully as it could save you from serious injury.

Always wear appropriate safety gear, including rubber-soled shoes, gloves (if working with cables, wires, or metal boxes); and safety glasses and a dust mask (when sawing or working overhead). And remember that current flows most easily along a path of least resistance, but it will follow any path that's available . . . including you!

Turn off the power to the circuit at the main service panel before removing receptacle, switch, or fixture covers.

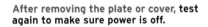

After removing the plate or cover, test again to make sure power is off.

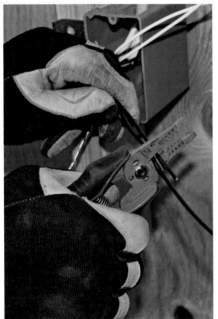

Wear gloves to protect your hands from the sharp edges of wires, cables, and metal boxes.

Wear safety glasses when sawing or drilling, especially when working overhead.

SERVICE PANELS

At the main service panel, the two hot cables from the meter base attach to lug terminals atop the main breaker. The incoming neutral cable attaches to the main lug of the neutral/ground bus. In main service panels, neutral/ground buses must be bonded, usually by a main bonding jumper. In subpanels and all other locations downstream from the main service panel, ground and neutral buses must be electrically isolated from each other.

In a main fuse box, the hot cables attach to the main power lugs, and the neutral cable to the main neutral lug. Whether the panel has breakers or fuses, metal buses run from the bottom of the main breaker/main fuse. Running down the middle of the panel, buses distribute power to the various branch circuits. Similarly, neutral/ground buses are long aluminum strips with many screws, to which ground and neutral wires attach.

Each fuse or breaker is rated at a specific number of amps, such as 15 or 20. When a circuit becomes overloaded, its current flow becomes excessively high. This causes its breaker to trip or its fuse strip to melt, thereby cutting voltage to the hot wires. All current produces heat; but as current doubles, the heat generated quadruples. If there were no breakers or fuses, current would continue flowing until the wires overheated and a fire started. Thus the amperage ratings of breakers and fuses are matched to the size (cross-sectional area) of the circuit wires.

The main breaker

All electricity entering a house goes through the main breaker, which is usually located at the top of a main panel. In an emergency, throw the main breaker switch to cut all power to the house. The main breaker is also

INSIDE THE SERVICE PANEL

- Incoming ground
- Hot feeder lines (incoming power)
- Incoming neutral
- Main breaker
- Hot bus bar (behind breakers)
- Neutral bus bar
- Ground bus bar

Branch circuits

The main panel houses incoming cables from the meter as well as the breakers and wires that distribute electricity to individual circuits. At the service, neutral conductors (white wires), equipment-grounding conductors (bare copper or green insulated wires), the metal service panel, and the grounding electrode system (grounding rods) must be bonded together.

- Hot cables from meter attach to lug terminals
- Main breaker
- Neutral/ground cable attaches here.
- Main bonding jumper
- Neutral/ground buses
- Hot buses
- Knockouts for circuit cables on all sides of panel

the primary overcurrent protection for the electrical system and is rated accordingly. (The rating is stamped on the breaker handle.) Thus if the main breaker for a 200-amp panel senses incoming current that exceeds its overload rating, the breaker will automatically trip and shut off all power.

Meter-main combos

Increasingly common are meter-main combos, which house a meter base and a main service panel in a single box. Meter-main combos allow a homeowner to put the main breaker outside the house, where it can be accessed in an emergency—say, if firefighters want to cut the power to a blazing house before they enter it. When the service panel is located outside, electricians typically locate the subpanel close by, inside the house, to minimize runs of large SER cable to the subpanel.

Fuse boxes

Breaker service panels are the most common type you'll find in today's homes, but many older homes still have fuse boxes. Fuses are among the earliest overcurrent devices, and they come as either Edison-type (screw-in) fuses or cartridge (slide-in) fuses. The Edison-style fuses that screw in like a light bulb are more common, and they have little windows that you can peer into to see if the filament is separated. Separation means that the circuit was overloaded and the fuse has blown. A blackened (from heat) interior could mean a short circuit—a potentially dangerous situation calling for the intervention of a licensed electrician. The less common cartridge fuses are used to control 240-volt (240v) circuits and are usually part of the main disconnect switch.

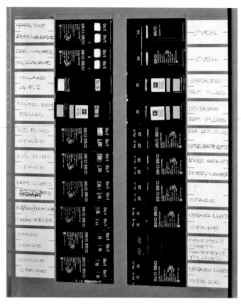

Labeling breakers speeds identification of the switch that turns off power to a device.

A meter-main combo, **placed outside the house, provides easy access for service or emergencies.**

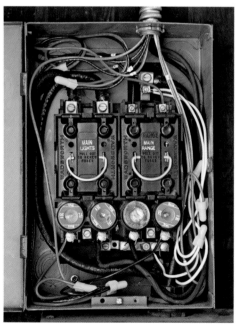

Stay away from the area **around the main breaker switch. It remains hot and is extremely dangerous**

A fuse box **in a subpanel.**

HOME ELECTRICAL SYSTEMS

Power from the utility service is commonly delivered through three large cables, or conductors, which may enter the house overhead or underground. Overhead service cables are called a *service drop*. These cables run to a weatherhead atop a length of rigid conduit. When fed underground, service conductors are installed in buried conduit or run as underground service-entrance (USE) cable. Whether it arrives overhead or underground, three-wire service delivers 240v.

Service conductors attach to a meter base and then to the service panel. Straddling the two sets of terminals on its base, the meter measures the wattage of electricity as it is consumed. The service panel also routes power to various circuits throughout the house.

The utility company will install cables to the building and will install the meter. The homeowner is responsible for everything beyond that, including the meter base and entrance panel, which a licensed electrician should install.

GFCI receptacles detect minuscule current leaks and shut off power almost instantaneously. They're important in kitchens and bathrooms, where water and dampness increase the potential for shocks.

Modern three-prong outlets **provide a low impedance path to ground in case a fault occurs.**

Some appliances require 30-amp service. This is a 125/250v dryer receptacle. The breaker for this circuit must also be rated for 30 amps.

⚠ WARNING

Whether the connection is an SER cable clamped to the lugs of a main breaker or between wires spliced together in an outlet box, the connection must be tight to be safe. Otherwise, electricity can leap a gap–it's called arcing–and that could lead to a house fire.

WIRING INSIDE OUT

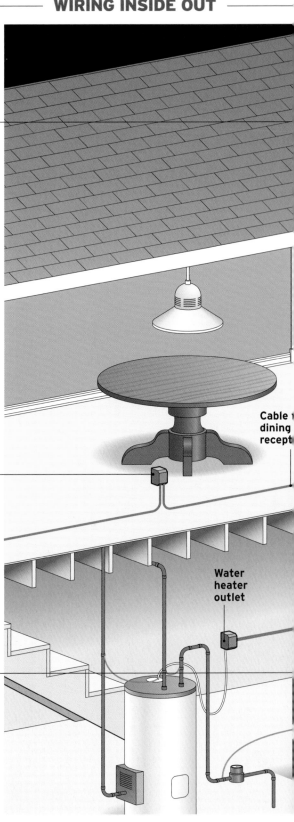

Cable
dining
recept

Water
heater
outlet

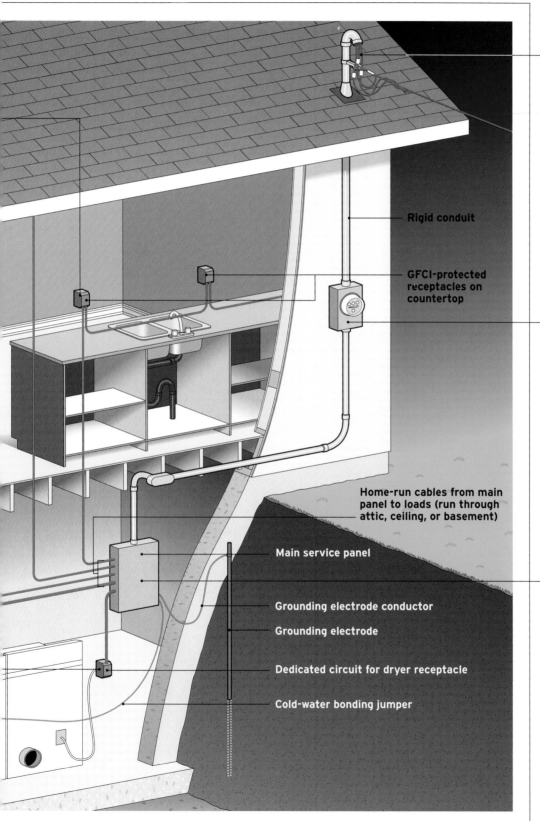

Rigid conduit

GFCI-protected receptacles on countertop

Home-run cables from main panel to loads (run through attic, ceiling, or basement)

Main service panel

Grounding electrode conductor

Grounding electrode

Dedicated circuit for dryer receptacle

Cold-water bonding jumper

A typical three-wire service assembly has two insulated hot conductors wrapped around a bare messenger cable, which also serves as the neutral.

The meter provides a measure of the electrical power consumed. Positioned outside the house, it allows the power company to monitor consumption.

The service panel distributes power to circuits throughout the house. Breakers interrupt power if the circuits become overloaded.

GROUNDING BASICS

Because electricity moves in a circuit, it will return to its source unless the path is interrupted. The return path is through the white neutral wires that bring current back to the main panel. In the event that the current seeks to return to the panel through a path other than the neutral wire, ground wires provide the current with an alternative low-resistance path.

Why is this important? Before equipment-grounding conductors (popularly called *ground wires*) were widespread, people could be electrocuted when they came in contact with fault currents that unintentionally energized the metal casing of a tool or an electrical appliance. Ground wires bond all electrical devices and potentially current-carrying metal surfaces. This bonding creates a path with such low impedance (resistance) that fault currents zip along it as they return to the power source, quickly tripping breakers or fuses and interrupting power. Contrary to popular misconceptions,

the human body has a relatively high impedance (compared to copper wire), so if electricity is offered a path with less resistance (a copper ground wire), it will take it.

➡ See "Avoiding Electrical Shocks," below.

Individual ground wires connect to every part of the electrical system that could become a potential conductor—metal boxes, receptacles, fixtures—and, through three-pronged plugs, the metallic covers and frames of tools and appliances. The connections, usually bare copper or green insulated wire, create an effective path back to the main service panel in case the equipment becomes energized.

The neutral/ground bus

The ground wires attach to a neutral/ground bus bar, which is itself bonded to the metal panel via a *main bonding jumper*. If there's a ground fault in the house, the main bonding

The smaller copper wire **at the top runs to a ground rod; the thicker copper wire below and the neutral wire feed a subpanel within the house.**

The main ground wire **from the service panel clamps to an 8-ft. rod-grounding electrode (or ground rod) driven into the earth. It diverts outside voltage, such as lightning strikes.**

Connections to cold-water pipes **prevent shocks should the pipes become inadvertently energized.**

AVOIDING ELECTRICAL SHOCKS

GROUND FAULTS CAN KILL

Current flowing through an unintended conductor, such as a person, is called a *ground fault*. **Because only a little current flowing through your heart can kill you, ground faults can be fatal. Likewise, you can get badly shocked by touching an energized wire or device with one hand and touching the other hand to a neutral or ground wire. In this case, you become part of the circuit and current flows through you.**

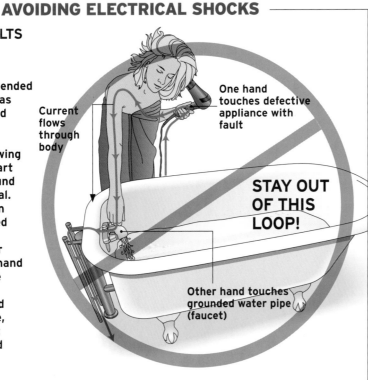

Current flows through body

One hand touches defective appliance with fault

STAY OUT OF THIS LOOP!

Other hand touches grounded water pipe (faucet)

jumper will ensure that the current can be safely directed to the ground—away from the house and the people inside. It is probably the single most important connection in the entire electrical system.

Also attached to the neutral/ground bus in the service panel is a large, bare copper ground wire—the grounding electrode conductor (GEC)—that clamps to a grounding electrode (also called a *ground rod*)—which is driven into the earth or attached to steel rebar in the footing of a foundation. The electrode's primary function is to divert lightning and other outside high voltages before they can damage the building's electrical system. Although the grounding electrode system (GES) is connected to the equipment grounding system at the service panel, the GES has virtually nothing to do with reducing shock hazards.

The National Electrical Code (NEC) sizes grounding electrode conductors based on the sizes and types of conductors in the service. Typically, residential GECs are size 6 American wire gauge (6AWG) copper. Ground rods are typically $5/8$-in. to $3/4$-in. copper-clad steel rods 8 ft. to 10 ft. long; the longer the rod, the more likely it will reach moist soil, whose resistance is less than that of dry soil. Be sure to install multiple rod systems in lightning-prone areas.

GFCIs

Ground-fault circuit interrupters (GFCIs) are sensitive devices that can detect even small current leaks and shut off power almost instantaneously. The NEC now requires GFCI protection on all bathroom receptacles; kitchen receptacles within 4 ft. of a sink; all receptacles serving kitchen counters; all outdoor receptacles; all accessible basement or garage receptacles; and all receptacles near pools, hot tubs, and the like.

➤ See "Wiring a GFCI Receptacle," on p. 48

MAJOR GROUNDING ELEMENTS

The equipment-grounding system acts as an expressway for stray current. By bonding conductors or potential conductors, the system provides a low-resistance path for fault currents. The abnormally high amperage (current flow) that results trips a breaker or blows a fuse, disconnecting power to the circuit.

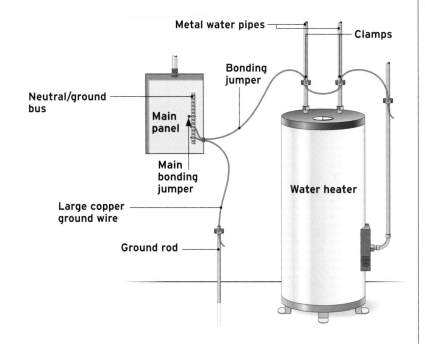

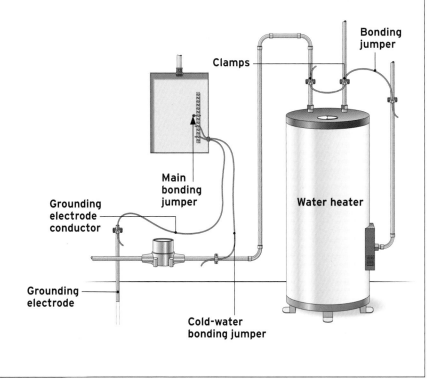

EVALUATING THE ELECTRICAL SYSTEM

Before you start working on your electrical system, you should have a general knowledge of its condition. Certain situations could have a big impact on your safety as you work. Others, such as the electrical requirements of your home, could affect whether you can add new circuits or certain kinds of devices. For your peace of mind and safety, you may want to hire a licensed electrician or qualified home inspector to advise you about what needs to be done. Start by considering the following aspects of your existing service.

The service

Is there two-wire or three-wire service to the house? In older homes, electrical service is often undersize for the demands of a modern household. If your house has only two cables running from the utility pole—one hot and one grounded neutral conductor—it has only 120v service and should be upgraded to three-wire, 240v service. A 100-amp, circuit-breaker service panel is considered minimal today.

General condition of panels

If you see scorch marks, rust stains, wires with frayed or cracked insulation, or condensation on the service panel or damp conditions around it, the service is unsafe. Dampness is particularly unsafe, and many electricians will refuse to work on a panel

This panel is so overloaded **that it will be tough to replace the cover.**

Sloppy, unstapled cable **can work loose or be damaged.**

These fuses are too large **(30 amps) for the load rating of the circuit wires. The wires could melt and start a fire.**

These scorch marks **indicate electrical arcing (electrical leaps between connections that have worked loose).**

Aluminum wiring Widely used in house circuits in the 1960s and 1970s, aluminum wiring expands and contracts excessively, which leads to loose connections, arcing, overheating, and—in many cases—house fires. The most common symptoms will be receptacle or switch cover plates that are warm to the touch, flickering lights, and an odd smell around electrical outlets. Once arcing begins, wire insulation deteriorates quickly. An electrician who checks the wiring may recommend adding COPALUM® connectors, CO/ALR-rated outlets and switches, or replacing the whole system. Aluminum service cable, however, is not a problem when terminated correctly and is still used today.

until the surrounding dampness is remedied. Look for melted plastic around wires, which could indicate overheating. Make sure the wire gauges match up with the size of the breaker (#14 wire on a 15-amp circuit, #12 on a 20-amp, and so forth). If the panel seems so stuffed with wires that you can barely close it, you should call a pro to do an inspection.

Current usage

Installing fuses too big for a circuit to prevent blown fuses is a fool's bargain; such circuits could overheat and start a fire. Make sure that heavy-duty circuits for ranges, dryers, and air-conditioners have the appropriately sized breakers installed. Likewise note over- loaded receptacles, extension cords under rugs, and the like inside the house. These are invitations for a fire and a definite sign that more receptacles are needed in the area.

Grounding

Is the main service panel grounded? There should be a large grounding wire running from the panel and clamped to a cold-water pipe and/or a grounding rod.

Is there equipment grounding? Grounding the panel is not enough. For the entire elec- trical system to be grounded, there must be continuous ground wires running to every device in the house.

If the house has only two-slot receptacles, the system may be grounded by wire-clad cables. Test the receptacles by inserting the probe of a voltage tester into the short slot and touch the other to the mounting screw. (Make sure it's not covered with paint.) If the tester lights, you can install a grounded receptacle. Otherwise, consider upgrading the circuit. Note that three-hole adapters can be used only in grounded receptacles. If there are three-slot receptacles, use a receptacle analyzer to check whether the plugs are grounded and polarized.

The ground is clamped **to the wrong side of a dielectric union (coupling) on this cop- per cold-water pipe, so there's no ground.**

All wire splices must be **protected by cov- ered boxes, and all electrical equipment must be installed in dry areas.**

Moisture wicking through **this masonry surface has rusted the fuse box and compromised electrical safety.**

A 30-amp main is inadequate **for modern uses; 200-amp service would be more appropriate.**

Arcing

When nails puncture wires or electrical con- nections are loose or corroded, electricity can arc (jump) between points. Arcing may be responsible for many of the 40,000 elec- trical fires each year. The NEC now requires arc-fault circuit interrupter (AFCI) protection for all 15-amp and 20-amp bedroom circuits. AFCIs detect minute fluctuations in current associated with arcing and de-energize the circuits before a fire can start. Installing AFCI breakers is essentially the same as installing GFCI circuit breakers.

CUTTING POWER AT THE PANEL

A In a normal breaker box, simply flip the relevant breaker to the off position.

B If you're working on a fuse panel, unscrew the fuse you're working on.

C If cartridge fuses are used, remove the cartridge or cartridge block.

D Once the power is off, post a warning sign to prevent others from turning the power back on while you're still working on it.

Always shut off the power to an outlet before working on it—and then test with an electrical tester to be sure there's no current present. Because individual devices such as receptacles, switches, and fixtures can give false readings if they are defective or incorrectly wired, the safest way to shut off the electricity is by flipping a breaker in the service panel or subpanel.

Turning off the power at a breaker panel is usually straightforward. After identifying the switch controlling the circuit, push the breaker's switch into the *off* position **A**. The breaker switch should click loudly into place: If it doesn't, flip it again until you hear a loud click. (A breaker that won't snap into place may be worn out or defective and should be replaced by an electrician.)

If your home has a fuse panel instead, remove the fuse **B** that controls the circuit, and tape a warning sign to the panel cover. Partially unscrewing a fuse is not a solution, because the fuse body is still in contact with the socket and possibly could be jiggled (or rescrewed) so that current resumes flowing through it. Remove the fuse. Likewise, if circuits are controlled by cartridge fuses **C**, pull the cartridge or cartridge block out of the panel.

In any case, once you've cut off the power, shut the panel cover and tape a sign to it, telling others to stay out **D**. Otherwise, someone not aware of the situation could flip the switch on, energizing the circuit you're working on.

TESTING WITH A NEON TESTER

When using a neon tester, hold the insulated part of the tester, and insert its probes into the receptacle slots. To test a three-slot receptacle, first insert the probes into the hot and neutral slots ❶. If the tester light does not glow, insert one probe into the hot (shorter) slot and the other probe into the round ground slot. If there's current present, the tester will light up ❷. Next, insert one probe into the neutral (longer) slot and the other into the ground slot. Here, the tester will glow only if there is current and the hot and neutral wires have been reversed ❸.

If the tester does not light up for any of these three combinations, there's probably no voltage present. However, if the receptacle is faulty or a wire is loose, the previous tests may not detect voltage present in the outlet box. To be certain, unscrew the receptacle cover plate and the two screws holding the receptacle to the outlet box. Pull the receptacle out from the outlet box, being careful not to touch bare wires, receptacle screw terminals, or metal outlet boxes.

Then apply tester probes to the bare wire ends on both sides of the receptacle and to the gold and silver screw terminals ❹. Finally, if the outlet box is metal, touch one probe to the hot wire/screw and the other probe to the metal box ❺. If the tester doesn't glow during these tests, the power is off.

1 Insert probes into both the hot and the neutral slots on the receptacle.

2 Test the hot (shorter) and the ground slots.

3 Test the neutral (longer) and the ground slots.

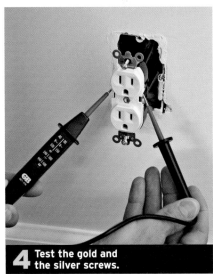

4 Test the gold and the silver screws.

5 Touch one probe to the hot wire/screw and the other to the metal box.

WARNING
To avoid shock, never touch the bare wire tips of the tester probes.

WARNING
Test the tester first. No matter what kind of tester you're using, test it on a circuit that you know is hot to make sure the tester is working properly.

TESTING WITH AN INDUCTANCE TESTER

An inductance tester is a popular battery-operated voltage tester that is reliable, inexpensive, and fits in a shirt pocket. Its tip glows when touched to a hot (energized) screw terminal or receptacle slot. Each time you use an inductance tester—or any voltage tester—first test its accuracy on a receptacle that you know is hot **1**.

After shutting off the power at the panel, first insert the tester tip into the short (hot) slot of a receptacle **2**. If the tester tip does not glow, there is probably no voltage present. To be sure, next insert the tester tip into the receptacle's long (neutral) slot **3**. This second insertion should protect you in case the receptacle was incorrectly wired.

If you need to remove the receptacle—say, to replace it—remove its cover. Then, being careful not to touch the sides of the receptacle, unscrew the two mounting screws holding the receptacle to the outlet box. (If the box is metal, avoid touching it, too.) Grasp the mounting straps and gently pull the receptacle out of the box. First touch the tip of the tester to the hot (black) wire or the gold screw terminal **4**, then touch the tip to the white (neutral) wire or silver screw terminal **5**. If the tester tip does not glow, it's safe to handle the receptacle and the wires feeding it.

An inductance tester will often glow when its tip is merely near a hot wire—that is, it can "read" current through a wire's insulation. Thus you can sometimes detect electrical current at a switch or fixture without removing the outlet cover **6**.

> ➡ **For more on testing switches and light fixtures, see Chapters 3 and 4.**

TRADE SECRET
Using an inductance tester is far safer than using a two-prong neon tester because it's possible to get shocked by touching the bare probes of neon testers.

1 Test the tester on an outlet you know to be hot.

4 Test the hot connector.

2 When checking to make sure power is off, test the hot slot first.

3 Then try the neutral slot.

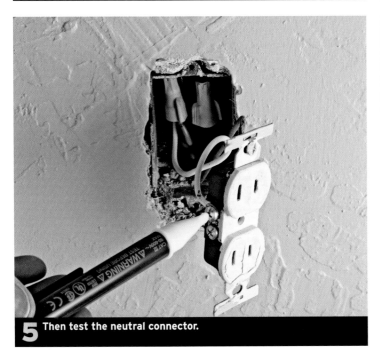

5 Then test the neutral connector.

6 You can often detect current through an outlet cover.

TOOLS & MATERIALS

YOU DON'T NEED A LOT OF expensive tools to wire a house successfully. And there's little uniformity among the tools electricians prefer. Some pros carry a dozen different pliers and wire strippers in their tool belts, whereas others streamline their movements and save time by using the fewest tools possible. This chapter introduces the basic tools and materials you'll need and a few of the basic techniques you'll perform repeatedly. All materials should bear the Underwriters' Laboratory (UL) stamp, which indicates that a component meets the safety standards of the electrical industry.

The first test of any tool is to fit your hand comfortably; the second, that it feel solid and well made. Better tools tend to be a bit heftier and cost more.

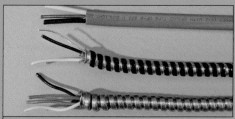

TOOLS

ELECTRICAL BOXES

CABLES & WIRES

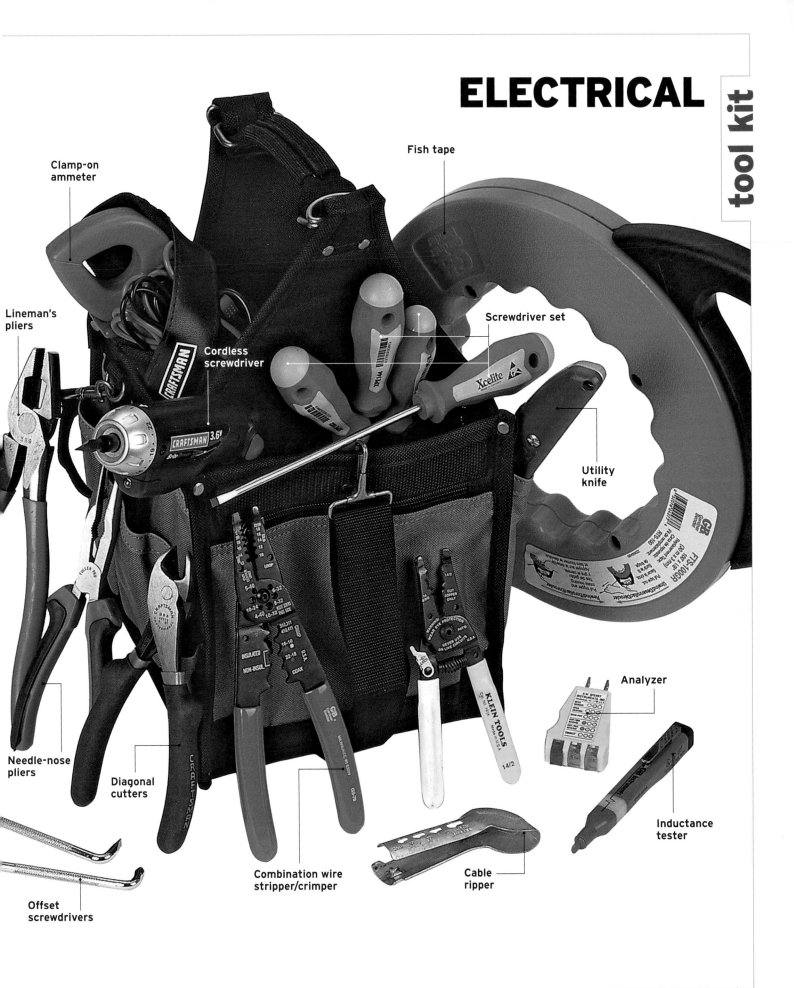

ELECTRICAL

Clamp-on ammeter

Fish tape

Lineman's pliers

Cordless screwdriver

Screwdriver set

Utility knife

Needle-nose pliers

Diagonal cutters

Combination wire stripper/crimper

Cable ripper

Analyzer

Inductance tester

Offset screwdrivers

HAND TOOLS

A tool belt is especially helpful for wiring work. Assign each tool a specific place in the belt and you can easily retrieve it without searching.

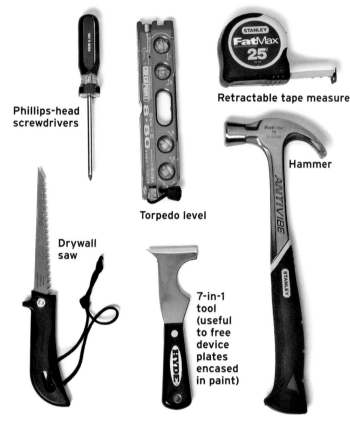

Phillips-head screwdrivers

Retractable tape measure

Torpedo level

Hammer

Drywall saw

7-in-1 tool (useful to free device plates encased in paint)

A ll hand tools should have insulated handles and fit your hand comfortably. Manufacturers now make tools in various sizes; to fit everyone from the largest male to the most petite female. Here again, don't scrimp on quality.

Pliers and strippers

Lineman's pliers are the workhorse of an electrician's toolbox. They can cut wire, hold wires fast as you splice them, and twist out box knockouts. *Needle-nose* (long-nose)

A comination stripping tool **not only cuts and strips wire but also crimps connectors, cuts small machine screws, and more.**

pliers can grasp and pull wire in tight spaces. These pliers can loop wire to fit around receptacle and switch screws. A large pair can also loosen and remove knockouts in metal outlet boxes. *Diagonal-cutting* and *end-cutting* pliers can cut wires close in tight spaces; *end cutters* (sometimes called nippers) also pull out staples easily. A *multipurpose tool* is used to strip individual wires of insulation, cut wire, crimp connections, and quickly loop wire around screw terminals.

A cable ripper strips the plastic sheathing from Romex® cable without harming the insulation on the individual wires inside. Many pros use a utility knife to strip sheathing, but that takes practice and a light touch to avoid nicking the insulation of individual wires. To strip

armored cable, use a *Roto-Split® cable stripper*; it's vastly superior to the old method of using a hacksaw and diagonal cutters.

Other useful tools

No two electricians' tool belts will look the same, but most contain a tape measure, flashlight, small level, hammer, Speed® Square, and a large felt-tipped marker. In the course of a wiring job, you may need several sizes of slot-head and Phillips®-head screwdrivers, plus an offset screwdriver and a nut driver.

Fish tape is used to run cable behind finish surfaces and pull wire through conduit. A fish tape is invoked in almost every old wiring how-to book on the market. Modern-day pros, however, swear by a pulling grip, also called a swivel kellum (p. 132). For remodel work, you may need a plaster chisel, flat bar, and a drywall saw.

POWER TOOLS

Buy power tools that are appropriate to your strength and to the task at hand. More powerful tools tend to be heavier and harder to manage; and for wiring, they're often overkill. If possible, test-drive a friend's power tool before buying your own.

Drills

A 1/2-in. right-angle drill allows you to fit the drill head between studs or joists and drill perpendicular to the face of the lumber.

> **To see a right-angle drill in action, see p. 177.**

Buy a drill with a clutch. Unless there's a clutch to disengage the motor, a drill bit that suddenly jams or lodges against a nail shank, could cause the body of the tool to torque powerfully and thus could injure you.

Drill bits

Spade bits cut quickly, but tend to snap in hard wood. For this reason most electricians rarely use spade bits. Self-feeding chipper bits drill doggedly through hard, old wood but won't last long if they hit nails. A 7/8-in. Greenlee® Nail Eater™ bit is your best bet if old lumber is nail infested.

Reciprocating saw

A reciprocating saw is indispensable for most remodeling jobs, whether to cut box openings in plaster or to notch framing. Choose a blade that's appropriate for the material you're cutting: coarser teeth for cutting wood, finer for plaster and metal. Special bimetal, remodel blades can cut through occasional nails without destroying the blade.

Rotary cutter

Using slightly different bits, a rotary cutter can cut through plaster or drywall to create box openings. Typically, drywallers install panels over installed outlet boxes, then use a rotary cutter to trim around the outside of a box; this method is much quicker than hand-cutting openings beforehand. Set the tools depth to avoid nicking wires.

Cordless power tools

Cordless drills and saws enable you to keep working when the powers is off or not yet connected. They don't need an extension cord and won't electrocute you if you inadvertently drill or cut into a live wire. Cordless reciprocating saws can cut anything from plaster lath to studs.

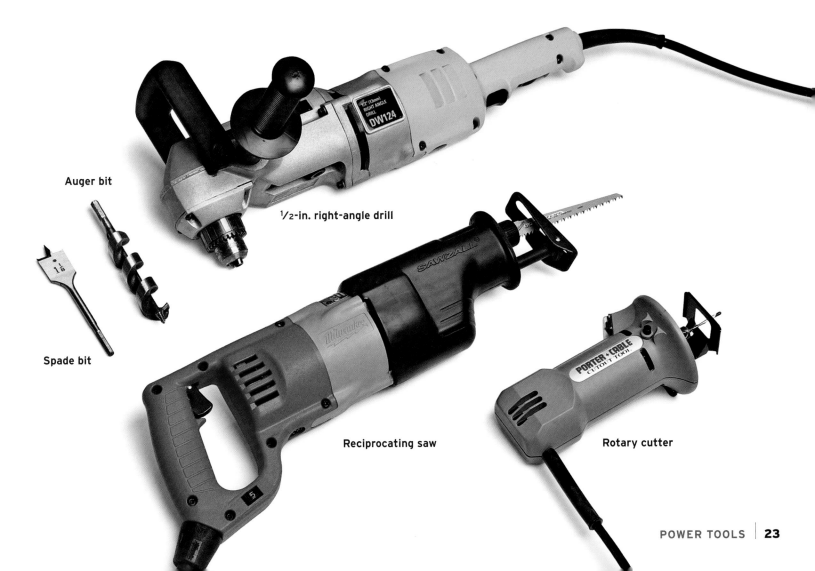

Auger bit

Spade bit

1/2-in. right-angle drill

Reciprocating saw

Rotary cutter

ELECTRICAL TESTERS

Testing to see if a circuit or device is energized is crucial to safety and correct wiring. There are several to choose from, and some perform multiple functions.

Neon voltage testers

Neon voltage testers are inexpensive and widely available, but there's a danger of touching the bare metal probes and getting shocked. Better neon testers have insulated handles. To use this tool, insert the probes into the receptacle slots or touch them to the screw terminals or to a metal outlet box to see if the unit is hot (energized).

Plug-in circuit analyzers

Plug-in circuit analyzers can be used only with three-hole receptacles, but they quickly tell you if a circuit is correctly grounded and, if not, what the problem is. Different light combinations on the tester indicate various wiring problems, such as no ground and hot and neutral reversed. They're quite handy for quick home inspections.

Solenoid voltage testers

Solenoid voltage testers (often called *wiggies*) test polarity as well as AC voltage. They also test DC voltage from 100v to 600v. Most models vibrate and light a bulb when current is present. Solenoid testers don't use batteries, so readings can't be compromised by low battery power. However, because of their low impedance, solenoid testers will trip ground-fault circuit interrupters (GFCIs).

Inductance testers

Inductance detectors provide a reading without directly touching a conductor. They often allow you to detect electrical currents without having to remove cover plates and expose receptacles or switches. Touch the tool's tip to an outlet, a fixture screw, or an electrical cord. If the tip glows

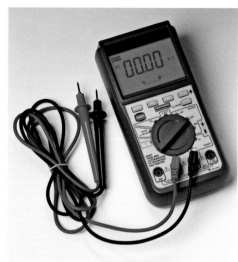

Some common electrical testers. A circuit ana-lyzer (top) shows problems in the wiring of a 3-pronged plug. Test whether an outlet is ener-gized with a neon voltage tester (left) or check for power with an inductance tester (right).

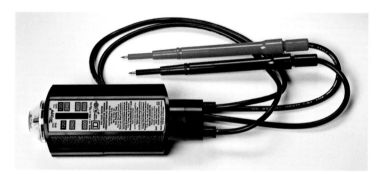

Digital multimeters provide precise read-ing in multiple scales.

Solenoid voltage testers don't require batteries to give a reading.

red, it means there's voltage present. Inductance testers rely on battery power.

Multimeters

A multimeter, as the name suggests, offers precise readings in multiple scales, which you select beforehand. Some models are even *autoranging*, meaning that they select the correct scale for you. Extremely sensi-tive, multimeters can detect minuscule amounts of voltage. Better models test AC and DC voltage, resistance, continuity, capacitance, and frequency.

Continuity tester

In addition to voltage testers, get a conti-nuity tester to test wire runs and connec-tors for short circuits or other wiring flaws. And be sure to do all your testing before a circuit is connected to power.

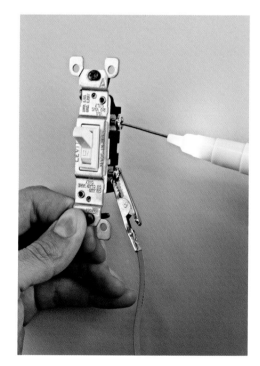

A continuity tester enables you to determine whether an electrical device or length of wire is defective.

CHOOSING ELECTRICAL BOXES

There is a huge selection of boxes, varying by size, shape, mounting device, and composition. But of all the variables to consider when choosing boxes, size (capacity) usually trumps the others. Correctly sized boxes are required by Code and faster to wire because you don't have to struggle to fit wires and devices.

Box capacity

The most common shape is a single-gang box. A single-gang box 3½ in. deep has a capacityof roughly 22½ cu. in.; enough space for a single device (receptacle or switch), three 12-2 w/grd cables, and two wire connectors. Double-gang boxes hold two devices; triple-gang boxes hold three devices. *Remember:* Everything that takes up space in a box must be accounted for—devices, cable wires, wire connectors, and cable clamps—so follow NEC recommendations for the maximum number of conductors per box.

You can get the capacity you need in a number of ways. Some pros install shallow four-squares (4 in. by 4 in. by 1½ in. deep) throughout a system because such boxes are versatile and roomy. If a location requires a single device, pros simply add a mud ring cover.

Because of their shallow depth, these boxes can also be installed back to back within a standard 2x4 wall. This allows you to keep even back-to-back switch boxes at the same height from one room to the next. Shallow pancake boxes (4 in. in diameter by ½ in. deep) are commonly used to flush mount light fixtures.

Metal vs. plastic boxes

Metal boxes are sturdy and are available in more sizes than are plastic boxes. Some metal boxes can be interlocked for larger capacity. Also, metal boxes are usually favored for mounting ceiling fixtures because steel is stronger than plastic. If Code requires steel conduit, armored cable (BX), or MC cable, you *must* use steel boxes. All metal boxes must be grounded.

>> >> >>

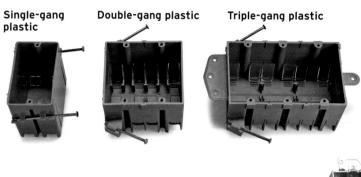

Single-gang plastic   Double-gang plastic   Triple-gang plastic

Single-gang adjustable with (orange) snap-on data ring   Double-gang adjustable   Single-gang metal

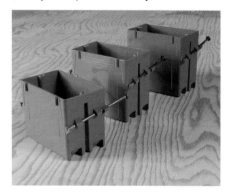

Single gang boxes come in three sizes, 18 cu. in., 20.4 cu. in. and 22.5 cu. in.

Throw a a single or double **gang mud (plaster) ring** on a 4-in. box and it's hard to overfill.

BOX FILL WORKSHEET*			
Item	Size (cu. in.)	Number	Total
#14 conductors exiting box	2.00		
#12 conductors exiting box	2.25		
#10 conductors exiting box	2.50		
#8 conductors exiting box	3.00		
#6 conductors exiting box	5.00		
Largest grounding device; count only one		1	
Devices; two times connected conductor size			
Internal clamps; one based on largest wire present		1	

*Table based on NEC 370-16(b) and adapted with permission from Redwood Kardon, Douglas Hansen, and Mike Casey, *Code Check: Electrical* (2005, The Taunton Press).

CHOOSING ELECTRICAL BOXES (CONTINUED)

For most other installations, plastic is king. (Plastic boxes may be PVC, fiberglass, or thermoset.) Electricians use far more plastic boxes because they are less expensive. Also, because they are nonconductive, they're quicker to install because they don't need to be grounded. However, even if a box doesn't need to be grounded, all electrical devices inside must be grounded by a continuous ground. Another reason to buy plastic: Box volumes are stamped on the outside.

Cut-in boxes

The renovator's mainstay are *cut-in boxes* because they mount directly to finish surfaces. These boxes are indispensable when you want to add a device but don't want to destroy a large section of a ceiling or wall to attach it to the framing. Most cut-in boxes have plastic ears that keep them from falling into the wall cavity; what vary are the tabs or mechanisms that hold them snug to the back side of the wall: screw-adjustable ears, metal-spring ears, swivel ears, or bendable metal tabs (Grip-Lok™ is one brand).

➡ **For information on installing remodel boxes, see p. 196.**

The screw on the side of an adjustable box **enables you to raise or lower the face of the box to make it flush to the finish wall.**

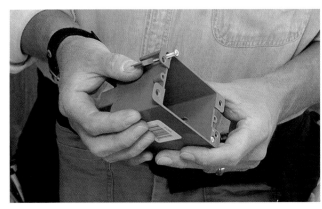

An old work box **doesn't mount to a stud. Instead, a pair of ears flips up at the turn of a screw and clamps the box to the wall.**

WARNING

All cut-in boxes, whether plastic or metal, must contain cable clamps inside that fasten cables securely. That is, it's impossible to staple cable to studs and joists when they are covered by finish surfaces, so you need clamps to keep the cables from getting tugged or chafed by the metal edge of the box.

CUT-IN BOXES

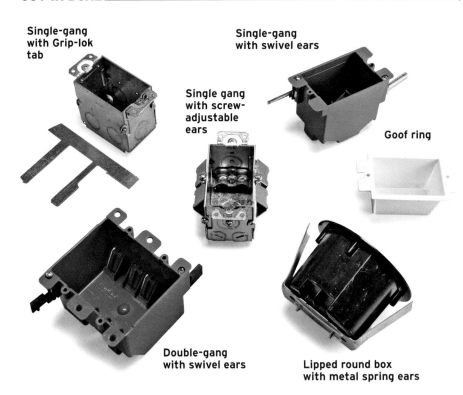

Single-gang with Grip-lok tab

Single-gang with swivel ears

Single gang with screw-adjustable ears

Goof ring

Double-gang with swivel ears

Lipped round box with metal spring ears

INSTALLING NEW WORK BOXES

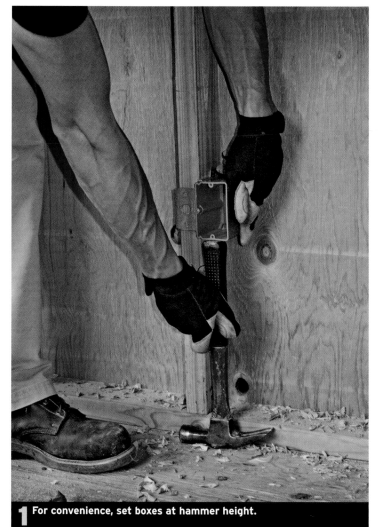

1 For convenience, set boxes at hammer height.

2 Screw an adjustable box to the framing.

There are few set rules about locating boxes. In general-use rooms, set the bottom of outlet boxes 12 in. above the subfloor—which you can approximate by resting a box atop a hammer held on end **1**. In housing for disabled occupants, outlet box bottoms should be a minimum of 15 in. above the subfloor.

For outlets over kitchen and bath counters, set box bottoms 42 in. from the subfloor, so they'll end up 8 in. above counters and 4 in. above a standard 4-in.-high back-splash. Locate wall switches on the lock side of a door (opposite the hinges).

▶ **For more about locating boxes, see p. 184.**

Mount boxes so that they'll be flush with finish surfaces. Most boxes have tabs or gauge marks stamped on the side to indicate different surface thicknesses. If that's not the case, hold a scrap of the finish material—for example, $5/8$-in. drywall—next to the front edge of the box as a depth gauge. Unless you're installing nail-in boxes, use screws to mount the boxes so you can make adjustments if you need to **2**. As noted elsewhere, adjustable boxes can be tweaked after the drywall is up.

(i) ACCORDING TO CODE

Before positioning outlet boxes, check to see if local building codes require them to be set at a certain height.

REMOVING KNOCKOUTS

Once you've mounted boxes, you'll need to remove the appropriate number of box knockouts and install cable connectors (clamps). Single-gang plastic boxes don't need clamps: Simply strike a screwdriver handle with the heel of your hand to drive out the knockout. To remove a metal-box knockout, jab it with the nose of needle-nose pliers to loosen it **A**, then use the plier's jaws to twist it free **B**.

Use a screwdriver **to remove a plastic-box knockout.**

A Strike a metal knockout to loosen it.

B Once the knockout is loose, remove it using pliers.

MOUNTING DEVICES

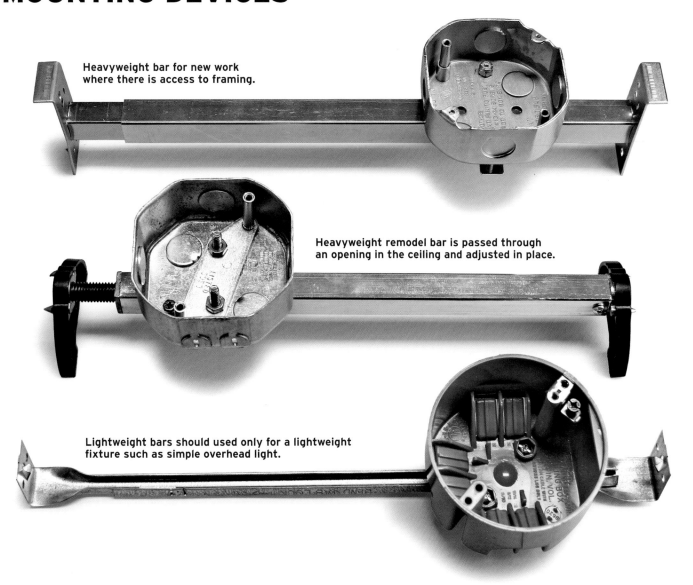

Heavyweight bar for new work where there is access to framing.

Heavyweight remodel bar is passed through an opening in the ceiling and adjusted in place.

Lightweight bars should used only for a lightweight fixture such as simple overhead light.

The type of mounting bracket, bar, or tab you use depends on whether you're mounting a box to finish surfaces or structural members. When you're attaching a box to an exposed stud or joist, you're engaged in new construction, or *new work* (as distinguished from *old work*, or remodel work), even if the house is old. New-work boxes are usually side-nailed or face-nailed through a bracket; nail-on boxes have integral nail holders. The mounting bracket for adjustable boxes is particularly ingenious. Once attached to framing, the box depth can be screw-adjusted until it's flush to the finish surface.

Adjustable bar hangers enable you to mount boxes between joists and studs; typically, hangers adjust from 14 in. to 22 in. Boxes mount to hangers via threaded posts or, more simply, by being screwed to the hangers. Bar hangers vary in thickness and strength, with heavier strap types favored to support ceiling fans and heavier fixtures.

A mounting bar for **new work** is screwed into the ceiling joists.

CABLE & CONDUIT

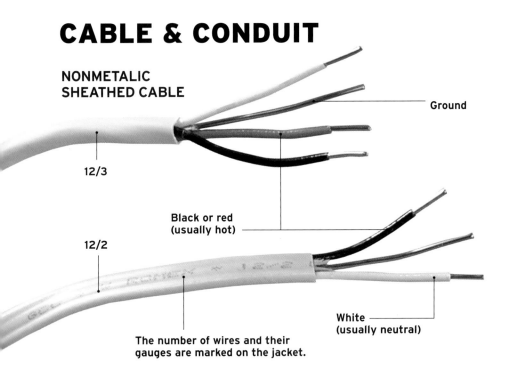

NONMETALIC SHEATHED CABLE

12/3

12/2

Ground

Black or red
(usually hot)

White
(usually neutral)

The number of wires and their
gauges are marked on the jacket.

Some old houses still have knob-and-tube
wiring, which doesn't necessarily need to be
replaced. Have it tested to make sure it's
still in good shape.

Most modern house wiring is flexible cable, but you may find any–or all– of the wiring types described here in older houses. Inside cables or conduits are individual wires, or conductors, that vary in thickness according to the load (amperage) they carry. Here's a quick overview.

Cable

Nonmetallic sheathed cable (NM or Romex) is by far the most common flexible cable. Covered with a flexible thermoplastic sheathing, Romex is easy to route, cut, and attach. Cable designations printed on the sheathing and the sheathing color indicate the gauge and the number of conducting wires inside.

➡ **See Reading a Cable, on the facing page.**

Most of the NM cable used for a standard 15-amp lights-and-outlets circuit is 14/2 w/grd. For a 20-amp circuit, use 12-gauge cable. Three-way switches are wired with 14/3 or 12/3. The third conductor is red. Either the red or the black wire can be hot, depending on the switch position.

➡ **For more on three- and four-way switches, see p. 51.**

Remember that you can wire 15-amp circuits with 12-gauge wire, but you can't use 14-gauge wire for 20-amp circuits. Metal-clad (MC) cable is often specified where wiring is exposed. Some codes still allow armored cable (AC), but that's increasingly rare.

Knob-and-tube

Knob-and-tube wiring is no longer installed, but there's still plenty of it in older houses. If its sheathing is intact and not cracked, it may still be serviceable. You may even be able to extend it, but have an electrician do the work. Knob and tube is eccentric, requiring experience and a skilled hand.

Circuit wiring

Copper is the preferred conductor for residential circuit wiring. Aluminum cable is frequently used at service entrances, but it is not recommended for branch circuits. Individual wires within a cable or conduit are color coded. White or light gray wires are neutral conductors. Black or red wires denote hot, or load-carrying, conductors.

Green or bare (uninsulated) wires are ground wires, which must be connected continuously throughout an electrical system.

Because most of the wiring in a residence is 120v service, most cables will have three wires: two insulated wires (one black and one white) plus a ground wire, usually uninsulated. Other colors are employed when a hookup calls for more than two wires–for example, 240v circuits and three- or four-way switches.

Conduit

Conduit may be specified to protect exposed wiring; it is commonly thin-wall steel (electrical metallic tubing; EMT), aluminum, or polyvinyl chloride (PVC) plastic. Metal conduit serves as its own ground. Apart from service entrances, conduit is seldom used in home wiring. When connected with weather-tight fittings and boxes, conduit can be installed outdoors.

➡ **For more on 240v circuits, see "Appliances", pp. 212–223.**

CABLE AND CONDUIT FOR SERVICE PANELS

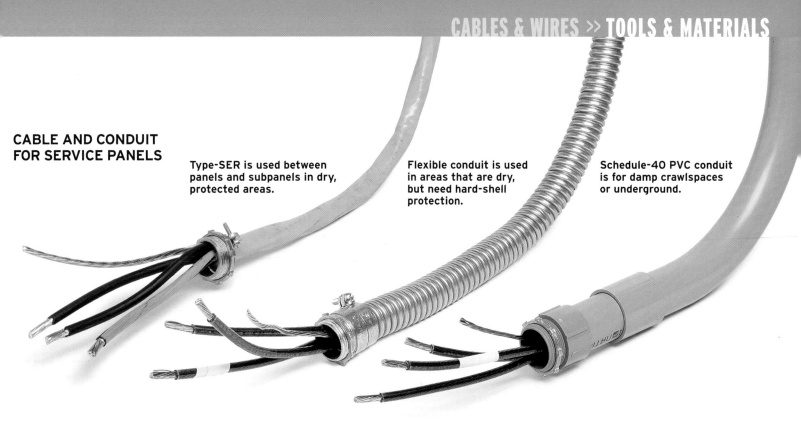

Type-SER is used between panels and subpanels in dry, protected areas.

Flexible conduit is used in areas that are dry, but need hard-shell protection.

Schedule-40 PVC conduit is for damp crawlspaces or underground.

READING A CABLE

Cables provide a lot of information in the abbreviations stamped into their sheathing—for example, *NM* indicates nonmetallic sheathing, and *UF* (for underground feeder) can be buried. The size and number of individual conductors inside a cable are also noted: *12/2 w/grd* or *12-2 W/G*, for example, indicates two insulated 12AWG wires plus a ground wire. Cable stamped *14/3 W/G* has three 14AWG wires plus a ground wire. (The higher the number, the smaller the wire diameter.) The maximum voltage, as in 600v, may also be indicated.

Individual wires within cable have codes, too. *T* (thermoplastic) wire is intended for dry, indoor use, and *W* means wet; thus *TW* wire can be used in dry and wet locations. *H* stands for heat resistant. *N*, for nylon jacketed, indicates a tough wire than can be drawn through conduit without being damaged.

Finally, make sure the cable is marked *NM-B*. Cable without the final *B* has an old-style insulation that is not as heat resistant as NM-B cable.

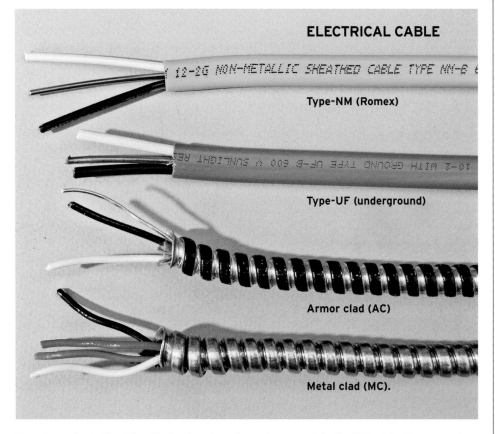

ELECTRICAL CABLE

12-2G NON-METALLIC SHEATHED CABLE TYPE NM-B

Type-NM (Romex)

10-2 WITH GROUND TYPE UF-B 600 V SUNLIGHT RE

Type-UF (underground)

Armor clad (AC)

Metal clad (MC).

The silver wire in the AC cable is a bonding wire, not a ground. In the MC cable, the green wire is ground, the white is neutral, and the red and black are hot.

CABLE CLAMPS & CLIPS

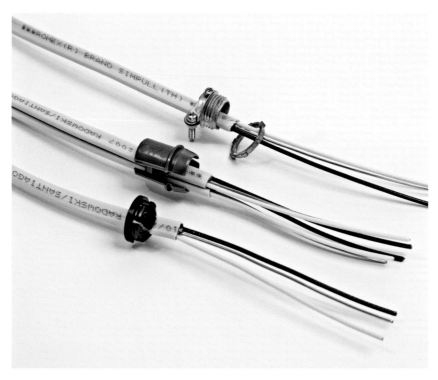

Romex cable connectors. From left: plastic push-in connector, two-cable hit-lock connector, 3/8-in. NM clamp with locknut, metal box with internal clamps. Cable connectors are set in box knockouts to prevent wires from wearing against sharp edges.

It's important to secure cable as you run it and to protect it from puncture or being accidently severed. It's also essential to make sure cable is tightly secured into boxes.

Clamps

Clamps secure cable to boxes to protect connections inside the box so wire splices or connections to devices cannot get yanked apart or otherwise compromised. Every wiring system—whether nonmetallic (Romex), MC, or conduit—has clamps (connectors) specific to that system. Cable clamps in metal boxes also keep wires from being nicked by burrs created when metal box knockouts are removed. (Use a screwdriver to start knockouts and lineman's pliers to twist them free—or a pair of heavy-duty needlenose pliers for both tasks.)

The exception to this rule is single-gang plastic boxes. If framing is exposed (new construction) and cable can be stapled within 12 in. of the box, Code doesn't require cable clamps in a single-gang plastic box. However, two-gang plastic boxes must have cable clamps— typically, a plastic tension clip that keeps cables from being pulled out. And, as noted in the preceding chapter, all cut-in boxes must contain cable clamps.

Two-piece locknut connectors **A** are still the most common cable clamp, but professional electricians who are racing the clock swear by plastic snap-in cable connectors **B**, which seat instantly and grip NM cable tight.

➡ For more on positioning boxes, see p. 27.

WARNING

Cordless drills and screwdrivers reduce the tedium of screwing wires to terminals, attaching devices to boxes, putting on cover plates, and connecting myriad other items. But always tighten cable clamps by hand to avoid overtightening them and damaging the incoming wires.

A A metal locknut connector consists of two separate pieces.

B Plastic snap-in connectors snap easily into place.

SECURING CABLE TO FRAMING

The quickest way to secure cable is to staple it. The trick is to staple it correctly—flat and not so tight that the sheathing is squeezed against the framing. Staples should be snug but not too tight. Use enough staples to secure cable, remembering that code requires staples at least every 54 in. Use particular care when stapling cable overhead ❶. Avoid making a sharp bend immediately after a staple, and leave yourself enough slack around boxes.

When a large number of cables run into a single box, it can be difficult to staple them within the 8 in. required by Code. Here's where cable clips come in handy. Simply nail or screw the clip to the stud where the box is attached ❷. Thread the cables into the clip for a neat, organized box ❸.

There are several styles of cable clips with different capacities. Avoid exceeding the capacity of the clip; it should hold the cable snugly but be loose enough to dissipate heat.

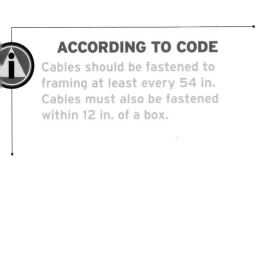

ACCORDING TO CODE

Cables should be fastened to framing at least every 54 in. Cables must also be fastened within 12 in. of a box.

1 Staple snugly but not tightly enough to squeeze the sheathing.

2 Nail cable clips within 12 in. of the box the cables will enter.

3 Neatly thread the cable into the clip.

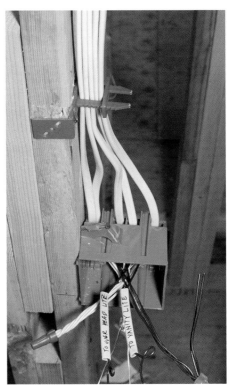

Another style of cable clip. It holds the cable near enough to the box but out of harm's way.

STRIPPING CABLE WITH A UTILITY KNIFE

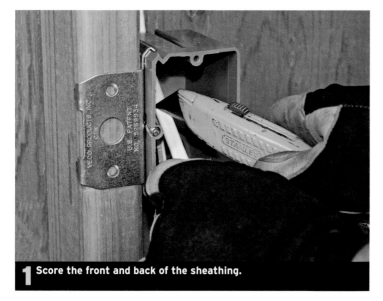

1 Score the front and back of the sheathing.

2 Slide the sheathing off of the cable.

3 Remove the kraft paper covering the ground wires, and you're ready to make connections.

Many electricians use a utility knife to slit and remove NM cable sheathing, but it requires a light touch that takes a lot of practice. Typically, pros hold the blade at a low angle to the cable and lightly run the blade tip down the middle of the cable and over the bare ground wire inside. Alternatively, one can score the front and back face of the sheathing with diagonal slits **1** and then yank the sheathing and slide it off **2**.

Once the sheathing is off, tear off the kraft paper covering the bare ground wires **3**. Then tuck the cables back into the box until you're ready to wire the receptacle or switch.

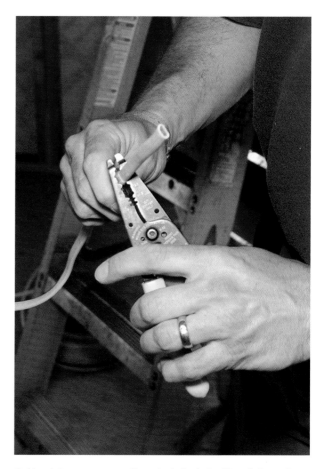

Cable strippers remove **the wire's jacket without damaging the conductor insulation.**

STRIPPING CABLE WITH A CABLE RIPPER

To remove sheathing with less risk of nicking wire insulation, use a cable ripper to slit the sheathing along its length ❶. Because the ripper's tooth is intentionally dull (so it won't nick wire insulation), it usually takes several pulls to slit the sheathing completely. Once that's done, pull back the sheathing and the kraft paper. You can snip both off using diagonal cutters ❷.

Because cable clamps grip sheathing—not individual wires—there should be at least ½ in. of sheathing still peeking out from under cable clamps when you're done. If you didn't tighten cable clamps earlier, do so now.

If there is only one cable entering a box, simply cut individual wires to length (typically, 8 in.) and tuck them into the box. If the box is metal, first bond the cable's ground wire to the box, using a grounding clip or a green grounding screw ❸. Once the wires are stripped and the box is grounded, fold the rest of the wires back into the box until you're ready to wire switches and outlets ❹.

1 Pull a cable ripper along the length of the last few inches of cable.

2 Snip both the sheathing and the paper.

3 Attach the ground wire to the metal box.

4 Tuck wires into the box until you're ready to wire the outlet.

CHOOSING WIRE

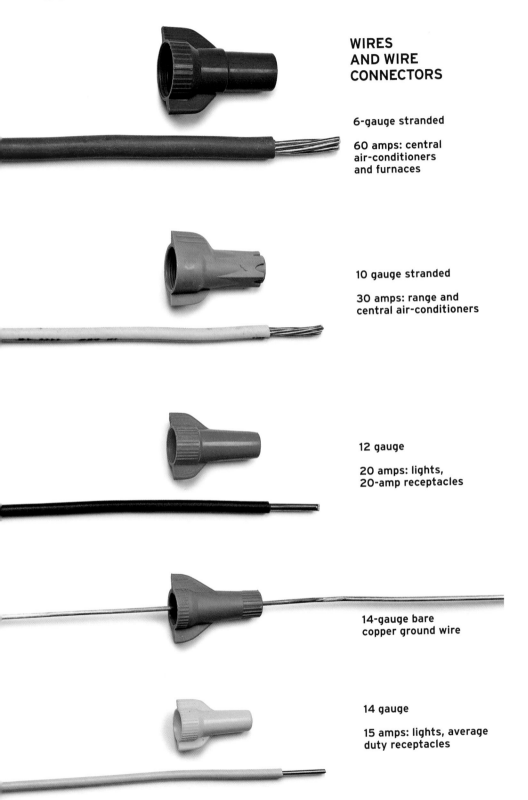

WIRES AND WIRE CONNECTORS

6-gauge stranded

60 amps: central air-conditioners and furnaces

10 gauge stranded

30 amps: range and central air-conditioners

12 gauge

20 amps: lights, 20-amp receptacles

14-gauge bare copper ground wire

14 gauge

15 amps: lights, average duty receptacles

Twist-on wire connectors are color coded to fit wires of different sizes. Green connectors, used to splice ground wires, have a hole in the cap that facilitates running a bare ground wire to a device or a metal box.

Wire comes in several gauges meant for different amp circuits. The higher the gauge, the smaller the wire. Larger wires can carry greater amperage, just as a larger pipe can carry greater water volume. If you use too small a wire, the resistance (measured in ohms) is too great and the wire can melt, causing a house fire. That's why it's important to use the right gauge wire for the load.

Wire connectors

Wire connectors, sometimes called by the popular brand Wire-Nut®, twist onto a group of like-colored wires to splice them together and ensure a solid mechanical connection. The importance of solid connections between spliced wires (or between wires and devices) can't be overstated. If wires work loose, electricity can leap the gaps between them and cause a house fire. Wire connectors are sized according to the number of wires and/or wire gauge they can accommodate; each size is color coded.

A divided pouch transforms a 5-gal. bucket into a portable hardware store of wire connectors, cable clamps, screws, staples, cable clamps, and other small items.

STRIPPING & SPLICING WIRES

Typically, electricians first splice the ground wires, which are usually bare copper. (If they're green insulated wires, first strip approximately 3/4 in. of insulation off their ends.) If you use standard wire connectors, trim the ground wires and butt their ends together, along with a 6-in. pigtail, which you'll connect later to the green ground screw of a receptacle. However, many pros prefer to twist the ground wires together, leave one ground long, and thread it through the hole in the end of a special wire connector ❶. Splicing hot and neutral wire groups is essentially the same. Trim hot wires to the same length ❷. Strip 3/4 in. of insulation off the cable wires and the pigtail ❸, and use lineman's pliers to twist the wires ❹. Then screw on a wire connector ❺.

When all the wire groups are spliced, gently fold the wires—rather like an accordion fold—and push them into the back of the box, where they'll be safe from drywall saws and nails ❻.

TRADE SECRET
Whenever you splice solid wires with a wire connector, twist the wires together before you twist the connector into place. This guarantees a solid connection between the wires should the wire connector come loose.

1 Twist ground wires and splice with a connector.

2 Trim hot wires, leaving enough length to work around the device.

3 Strip wire ends using wire stripper (approximately 3/4 in.)

4 Twist the wires together with lineman's pliers.

5 Twist on a wire connector until it's tight.

6 Carefully fold the wires into the box.

RECEPTACLES & SWITCHES

RECEPTACLES AND SWITCHES are the most-used electrical devices in a house. They're generally reliable and offer a long life, but they are often replaced when they become cracked or outdated or, eventually, cease to work. Fortunately, replacing them is straightforward and safe if you first shut off the power to the circuits that feed them—and test with a voltage tester to be sure that power is, in fact, off. This chapter shows you how to safely remove receptacles and switches, how to use a continuity tester to see if devices are defective, and how to install new receptacles or switches.

Wiring an electrical device is considered part of finish wiring—also called the trim-out stage—when finish walls are in place and painted. At the trim-out stage, everything should be ready so that the electrician needs only a pair of strippers and a screwdriver or screw gun.

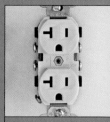

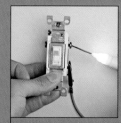

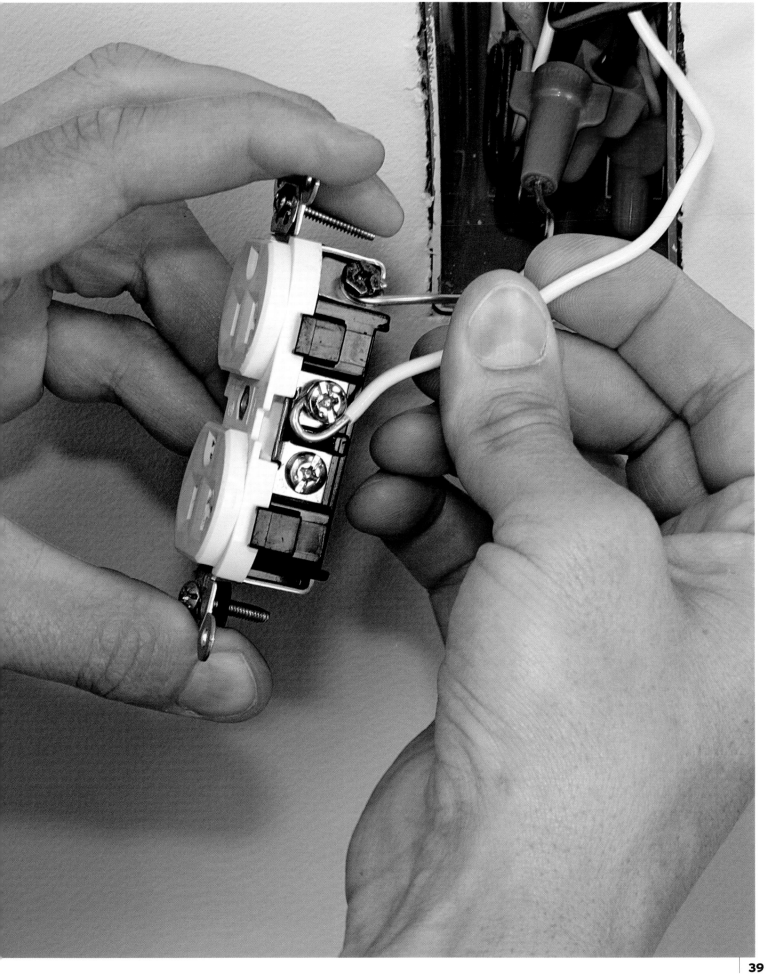

CHOOSING RECEPTACLES & SWITCHES

The difference in quality from one receptacle or switch to another can vary greatly. Over the life of the device, the difference in price is trivial, but the difference in performance can be substantial. For this reason, buy quality. As you can see in the top photo at right, cheap receptacles are pretty much all plastic, their thin metal mounting tabs will distort easily, and they tend to crack if subjected to heavy use.

On the other hand, quality receptacles and switches tend to have heavier nylon faces and may be reinforced with metal support yokes that reinforce the back of the devices.

Another telling detail is how wires are attached—whether they're screwed to terminals on the side of the device, inserted into the back of the device and held by internal clamps (back-wired), or some combination of the two. Again, better quality devices have better mechanisms for gripping wire.

For more on back-wired devices, see p. 42.

Most household receptacles are rated for 15-amp circuits and wired with 14AWG or 12AWG wire. The National Electric Code (NEC) specifies 20-amp protection for kitchen appliance, garage, and workshop circuits.

The NEC specifies ground-fault circuit interrupter (GFCI) protection for many locations, including bathroom, outdoor, and kitchen-counter receptacles, so there are also 15-amp and 20-amp GFCI receptacles. In addition, you can obtain various kinds of specialty 15-amp receptacles, including childproof models that require an adult's grip to uncover them, weatherproof receptacles that combine cover and receptacle in an integral unit, recessed receptacles in which clock wires can be hidden, and covered floor receptacles. In addition, there are receptacles specifically matched to the

ACCORDING TO CODE

All bathrooms and kitchens should have GFCI receptacles. All outdoor outlets and some garage outlets must also be GFCIs. Your local building code will have the final say on GFCIs.

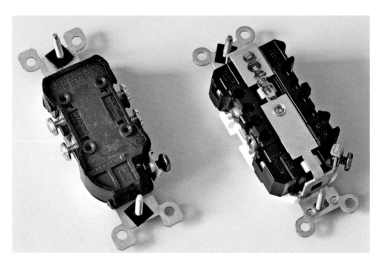

Better quality receptacles and switches **are usually heftier and more reliable. The quality receptacle on the right has a nylon face and its back is reinforced with a brass yoke.**

Switches (from left): **single-pole, three-way, four-way.**

Specialty switches (from left): **A timer switch, paddle-switch dimmer with small slide dimmer, linear slide dimmer.**

plugs of 30-amp, 40-amp, and 50-amp appliances. Your electrical supplier can help you find the right receptacle for your needs.

Matching load ratings

Circuit components must be matched according to their load ratings. That is, a 20-amp receptacle must be fed by 12AWG cable, which is also rated at 20-amps, and protected by a 20-amp breaker or fuse. A 15-amp receptacle or switch must be fed by 14AWG cable, which is rated for 15-amps, and protected by a 15-amp breaker or fuse.

Cable sheathing is color coded to help you correctly match wire size and devices: White sheathing denotes 14-gauge wire; yellow sheathing, 12-gauge; orange sheathing, 10-gauge; and so on.

Polarized receptacles

Receptacles, plugs, and fixtures are *polarized* so they can fit together only one way. A receptacle's gold screw terminal connects to hot wires and, internally, to the hot (narrow) prong of a polarized plug. The receptacle's silver screw terminal connects to neutral wires and, internally, to the neutral (wide) prong of a polarized plug. Finally, the green ground screw connects to the ground wire and the U-shaped grounding prong of the plug.

POLARIZED RECEPTACLES

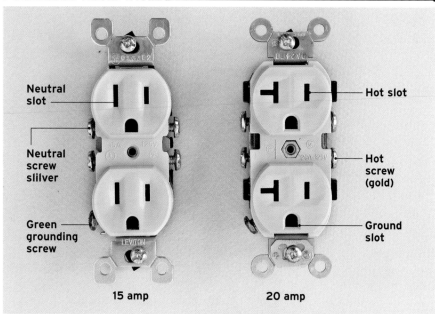

Neutral slot — Hot slot

Neutral screw sliiver — Hot screw (gold)

Green grounding screw — Ground slot

15 amp   20 amp

The 20-amp receptacle (at right) has a T-shaped neutral slot so it can receive a special 20-amp plug in addition to standard 15-amp plugs. But 15-amp receptacles cannot receive 20-amp plugs. Both receptacles are also polarized, so that only the large blade of a plug can fit into the large slot of the receptacle.

RECEPTACLES FOR DIFFERENT LOADS

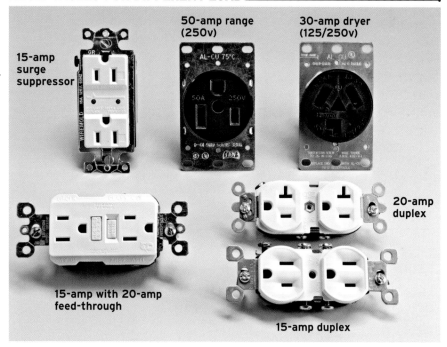

15-amp surge suppressor

50-amp range (250v)

30-amp dryer (125/250v)

20-amp duplex

15-amp with 20-amp feed-through

15-amp duplex

BACK-WIRED DEVICES

Back-wiring receptacles or switches is a faster alternative to wrapping wires around screw terminals. Back-wired devices have holes in the back, into which you insert stripped wire ends. But although back-wiring is quicker, many electricians—and some local electrical codes—feel it is unsafe. Their primary objection is that receptacles' internal tension clamps are made of thin metal strips, which can fatigue, leading to loose wires, flickering lights, and arcing. Moreover, each time users insert or remove plugs, the receptacles move slightly, which increases the likelihood of clamp failure. Manufacturers have stopped making back-wired 20-amp receptacles, so think twice about installing back-wired 15-amp receptacles.

Of course, there's an exception to every rule. A high-quality device such as the GFCI receptacle shown in photo ❶ on p. 48 allows you to loop wires around its screw terminals or insert wires into holes on the back of the device. Here, back-wiring is acceptable because you must screw down screws on either side to tighten internal clamps that grip the wire and ensure a solid connection.

Finally, back-wired switches are acceptable to Code. They rarely fail because switches aren't subject to the stresses of inserting and removing plugs; thus wire connections stay solid. Nonetheless, many pros don't like back-wired switches because their tension clamps can also fatigue and loosen. Use a screw terminal, they argue, and you're guaranteed a solid wire connection.

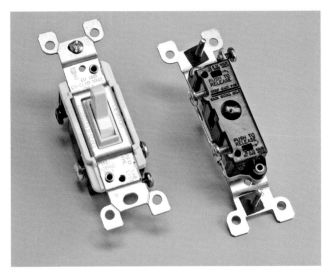

Many switches give installers **the option of connecting wires to screw terminals on the side or of back-feeding wires into holes in the back of the switch body.**

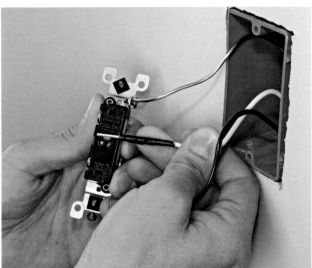

Back-wired switches are acceptable to most electrical codes. Use the stripping gauge on the back of the switch to determine how much insulation to strip from the wire.

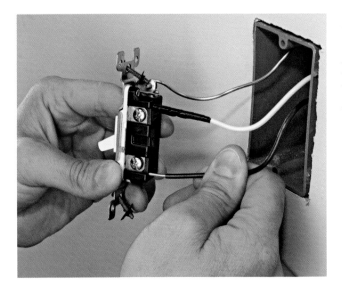

After stripping the wire end, **insert it into a back port/hole until the wire bottoms, then pull gently to make sure that the device's internal clamp has gripped the wire securely.**

TESTING FOR POWER

To identify the circuit that serves a particular receptacle, insert tester prongs into the receptacle and have a helper at the service panel flip breakers until the tester light goes out. To identify the circuit that serves a particular switch, turn on the fixture it controls and flip breakers until the light goes out. If that test is inconclusive or you aren't sure the receptacle or switch is operable, remove the cover plate and the two screws holding the device to the box. Being careful not to touch screw terminals or wires with your fingers, pull the receptacle out of the box. Touch tester prongs to each screw and to spliced wire groups. Here, an inductance tester is superior to a two-prong voltage tester because the inductance tester can usually read current through wire insulation or a wire connector.

WARNING

It's worth noting that in some old houses, the neutral wires—rather than the hot wires—may be attached (incorrectly) to receptacles or switches, in violation of Code. When testing existing receptacles, switches or fixtures, test *all* wires for voltage.

Before touching a wired receptacle, **switch, or fixture, use an inductance tester to see if power is present.**

TWO WAYS TO WIRE A RECEPTACLE

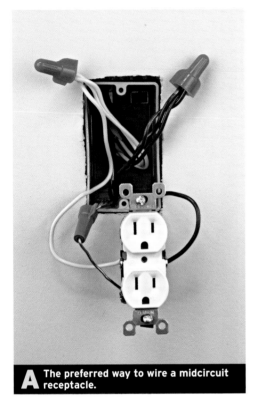

A The preferred way to wire a midcircuit receptacle.

B At the end of a circuit, wires from the cable attach directly to the receptacle.

C An alternate to wiring a midcut receptacle, without using pigtails. (Not recommended).

The duplex receptacle is the workhorse of house wiring, because it enables you to plug in a variety of energy users at locations around the house. Receptacles are so indispensable to modern life that Code dictates that no space along a wall in a habitable room should be more than 6 ft. from a receptacle and any wall at least 2 ft. wide must have a receptacle.

➤ **For more receptacle requirements, see. p. 40.**

The NEC requires that all conductors (wires) be solidly connected to devices such as receptacles, but several different wire configurations are acceptable. In brief, you can use pigtails—short lengths of wire running from wire splices to a device—or attach conductors directly to devices. Using pigtails to connect conductors ensures continuous power downstream, but sometimes it makes sense to connect conductors directly to the device.

Most electricians prefer using pigtails to connect conductors to midcircuit receptacles because it ensures continuous power and, if it's necessary to replace the receptacle at some future date, there are only three pigtails to disconnect. Wiring with a pigtail is shown in the photo **A**.

At the end of a circuit, where only one cable feeds an outlet, there's no need for pigtails. Just attach incoming wires directly to the receptacle as shown in photo **B**. As with pigtail wiring, connect the ground wire first, then the neutral, then the hot wire.

Route the circuit through the receptacle

Feeding circuits through receptacles is a widespread wiring method because, on the whole, it is quicker, requires fewer conductors, and results in boxes that are less crowded than those wired with spliced wires and pigtails **C**.

But this method has detractors, who argue that in a circuit so wired, if a receptacle upstream fails or a wire comes loose, receptacles downstream will lose power. For this reason, feeding a circuit through a receptacle is particularly risky if you also use back-wired receptacles whose internal clamping mechanisms can weaken and result in loose connections. Moreover, there is a voltage drop of about 1 percent per receptacle wired in this manner because receptacles offer more resistance to current flow than do wires. If you have 6 or 10 receptacles daisy-chained in this manner, those overheated connections can waste energy and increase your electricity bill.

➤ **See also "Back-Wired Devices," on p. 42.**

WIRING A DUPLEX RECEPTACLE

When a duplex receptacle is in the middle of a circuit, there will be two 12/2 or 14/2 cables entering the box—one from the power source and the other running downstream to the next outlet. In the sequence shown here, there are two 12/2 cables because the box houses a 20-amp receptacle.

To ensure continuity downstream, all wire groups will have been spliced with wire connectors during the rough-in stage. A pigtail from each splice will need to be connected to a screw terminal on the receptacle. Unless the small tab between screw pairs has been removed, you need attach only one conductor to each side of the receptacle.

Loop and install the ground wire to the receptacle's green grounding screw first. Place the loop clockwise on the screw shaft so that when the screw is tightened down, the screw head will grip—rather than dislodge—the wire ❶.

Next, loop and attach a neutral conductor to a silver screw terminal. Tighten down the screw that you don't use to avoid electromagnetic interference on radio receivers and the like. Then flip the receptacle over to access the brass screw terminals on the other side. If a looped wire end is too wide, use needle-nose pliers to close it ❷.

>> >> >>

1 Connect the ground wire first.

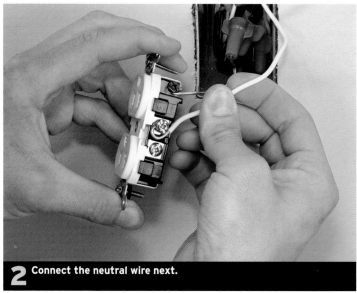

2 Connect the neutral wire next.

Wiring in an orderly way Any habit that increases your safety is worth adopting. When connecting wires to devices, most electricians connect the ground wire first, then the neutral wire, and then the hot wire. When disconnecting wires, they reverse the order: Disconnect the hot first, then the neutral, then the ground wire. Because the ground wire offers the lowest impedance path to ground, it makes sense to leave it connected as long as possible. Even if you're working on circuits that are disconnected, as veteran electricians say, "Treat every conductor as if it were live and you'll stay alive."

WIRING A DUPLEX RECEPTACLE (CONTINUED)

Screw down the brass screw so that it grips the hot wire. Pros frequently use screw guns for this operation, but weekend electricians should tighten the screw by hand to ensure a solid connection ❸.

Push the wired receptacle into the box by hand, keeping the receptacle face parallel to the wall ❹. Then hand-screw the device to the box. Avoid the temptation to use a screw gun because it can strip the screw holes in a plastic box ❺. Finally, install a cover plate to protect the electrical connections in the box and to prevent someone from inadvertently touching a bare wire end or the end of a screw terminal.

TRADE SECRET
It doesn't matter whether you install three-slot (grounded) receptacles with the ground slot up or down—just be consistent throughout the house.

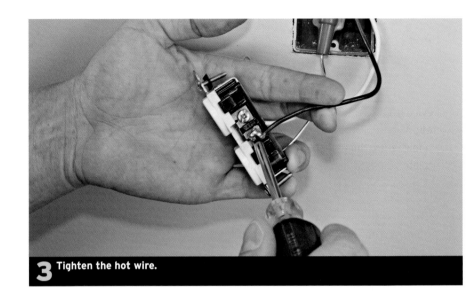

3 Tighten the hot wire.

4 Push the device into place by hand.

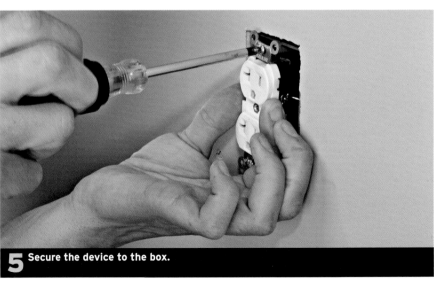

5 Secure the device to the box.

MAKING RECEPTACLE BOXES FLUSH TO DRYWALL

Outlet boxes are frequently installed below the surface of the drywall. If you're using adjustable boxes, that's not a problem because you can turn a screw to raise the box until it's flush with the drywall. It's imperative to bring the device flush to the drywall and to mount it securely. After a plug is inserted into the receptacle a few times, the receptacle moves and the cover plate cracks, which is both unsightly and unsafe.

If you're using a typical nail-on box, you can use plastic spacers (often called *caterpillars*) to build up the level of the receptacle so its mounting tab is flush to the drywall. These spacers take up the space between the mounting plate on the box and the device, so the mounting plate can be flush. Break off the pieces from the strip. This style folds. Insert the spacer behind the screw tabs.

TRADE SECRET

If a box is recessed more than ¼ in. from the surface, you must use a "goof ring" (p. 26).

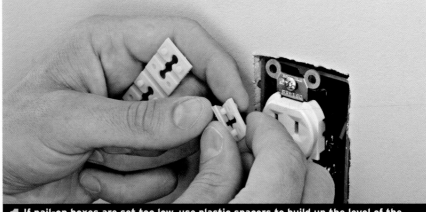

1 If nail-on boxes are set too low, use plastic spacers to build up the level of the receptacle or switch so its mounting tab is flush to the drywall.

2 After inserting the spacers between the device's mounting tab and the edge of the box, screw the device solidly to the box.

TWO-SLOT RECEPTACLES

Receptacles with two slots (instead of three) are ungrounded.

Receptacles with only two slots are ungrounded. Because they are fed by two-wire cable without a ground wire, they are inherently less safe than three-slot receptacles fed with a grounded cable. If existing cables and receptacles are correctly wired and in good condition, most codes allow you to keep using them. Should you add circuits, however, Code requires that they be wired with grounded cable (12/2 w/grd or 14/2 w/grd) and three-slot receptacles.

Replacing a two-slot receptacle with a GFCI receptacle can be a cost-effective way to add protection to a circuit. There will still not be a ground wire on the circuit, but the GFCI will trip and cut the power if it detects a ground fault. You can wire the GFCI receptacle to protect just that outlet or that outlet and all outlets downstream (away from the power source).

➡ **For more on wiring GFCI receptacles, see p. 48.**

Note: If one slot of a two-slot receptacle is longer, the receptacle will be polarized. That is, a receptacle's gold screw terminal will connect to a hot wire and, internally, to the hot (narrow) prong of a polarized two-prong plug. The receptacle's silver screw terminal connects to neutral wires and, internally, to the neutral (wide) prong of a polarized plug.

➡ **For more about polarity, see p. 41.**

WIRING A GFCI RECEPTACLE

When wiring a GFCI receptacle, it's important to connect incoming wires (from the power source) to the terminals marked "line" on the back of the receptacle. Attach outgoing wires (to outlets downstream) to terminals marked "load." To distinguish line and load wires during rough-in, write each term on small pieces of the cable sheathing and slip them over the appropriate wires before folding them into the box.

The GFCI shown here is something of a hybrid because it allows you to loop stripped wire ends around the screw terminals or to leave the stripped wire ends straight and insert them into holes in the back of the device—also known as back-wiring. In this case, back-wiring is acceptable because you must tighten screws on either side to engage internal clamps that grip the wire, thus ensuring a solid connection.

➤ For more on back-wired receptacles, see p. 42.

If the GFCI is going to protect users at a single outlet, attach wires to only one set of screw terminals ❶. The yellow tape across one set of screws indicates that they are load terminals: If you are hooking up the device to protect only a single point of use, leave the tape in place and connect wires only to the screw terminals marked "line." After attaching the ground pigtail, screw down the silver screw to secure the neutral pigtail.

Connect the hot pigtail to the brass screw last, then push the device into the box carefully, hand-screw it to the box, and install a cover plate ❷.

Feeding the circuit through a GFCI receptacle

If you want a GFCI receptacle to protect the outlet and all outlets downstream, feed the circuit through the receptacle. That is, connect incoming and outgoing cable wires directly to the device, rather than using pigtails. Again, it's important to connect incoming wires to the terminals marked "line," and outgoing wires to terminals marked "load."

➤ For more on wiring through the receptacle, see p. 44.

Quality GFCIs can be back-wired or side-wired.

1 A GFCI receptacle can be wired to protect a single outlet.

2 Connect the hot pigtail last.

or

Circuits can be fed through a GFCI receptacle.

WIRING A SPLIT-TAB RECEPTACLE

Standard duplex receptacles have a small metal tab between the brass screw terminals, which conducts power to both terminals, even if you connect only a hot wire to just one terminal. However, if you break off and remove that tab, you isolate the two terminals and create, in effect, two single receptacles—each of which requires a hot lead wire to supply power.

This technique, known as split-tab wiring, is often used to provide separate circuits from a single outlet, a configuration commonly used when connecting a disposal and a dishwasher. The disposal receptacle is almost always controlled by a switch, which allows you to turn off the disposal at another location. To supply two hot leads to a split-tab receptacle, electricians usually run a 12/3 or 14/3 cable.

> **For more on split-tab circuits, see the drawings on pp. 176 and 177.**

To create a split-tab receptacle, use needle-nose pliers to twist off the small metal tab between the brass screws ❶. Next connect the bare ground wire to the green grounding screw on the device and connect the white neutral wire to a silver screw. If you keep a slight tension on the wires as you tighten each screw, they'll be less likely to slip off ❷.

Flip the receptacle over to expose the brass screws on the other side, and connect a hot lead to each brass screw. If you're running 12/3 or 14/3 cable, one hot wire will typically be red and the other black ❸. Finally, push the device into the box by hand until it's flush and install the cover plate ❹.

Though this 15-amp split-tab receptacle is fed with 12/3 cable (rated for 20 amps), there's no danger of the load exceeding the rating of the receptacle. Because of the configuration of its slots, the receptacle can receive only a 15-amp plug.

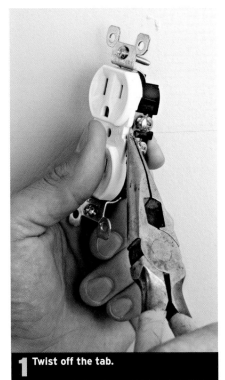

1 Twist off the tab.

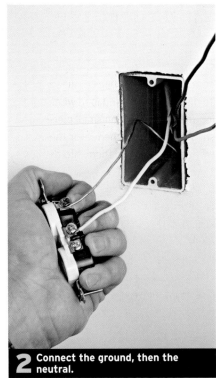

2 Connect the ground, then the neutral.

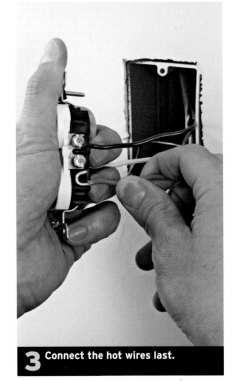

3 Connect the hot wires last.

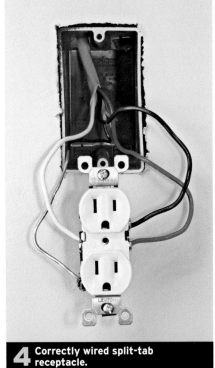

4 Correctly wired split-tab receptacle.

TESTING A SINGLE-POLE SWITCH

To test a disconnected switch, attach the tester clip to one screw or lead wire of the switch and touch the tester probe (point) to the other ❶. Turn the switch off and on ❷. If the tester doesn't light—or doesn't go off in either position—the switch is a dud. Replace it.

Because they're always racing the clock, many pros will simply replace a switch or receptacle that's suspect. But testing a device makes sense for homeowners. Continuity testers are inexpensive and easy to use. Testing a switch can save amateur electricians—who are more likely to miswire a device—a trip to the hardware store to replace a switch that isn't defective.

Note: Use continuity testers only on devices that are not connected to wiring.

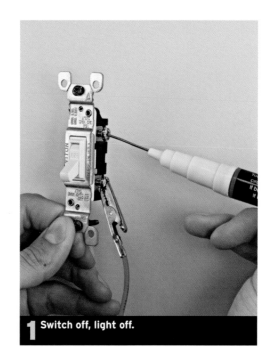

1 Switch off, light off.

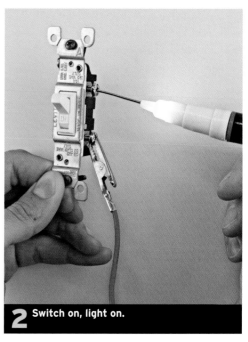

2 Switch on, light on.

Test the tester

Before you start, make sure the tester is working properly. Touch the point and the clip of the continuity tester to each other to be sure the tool is working. This completes the circuit and lights the tester bulb.

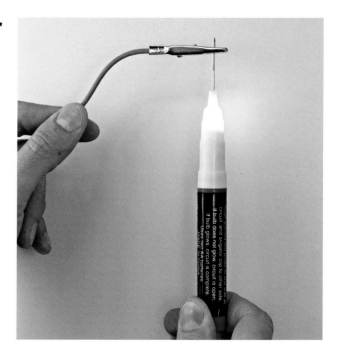

TESTING A THREE-WAY SWITCH

When testing a three-way switch, attach the tester clip to the common terminal and touch the probe to one of the traveler screw terminals. The tester should light ❶. If the switch is functioning properly, flip the switch and the light should turn off ❷. Leave the clip in place, touch the probe to the other traveler screw, and flip the switch again. If all that works, the switch is good.

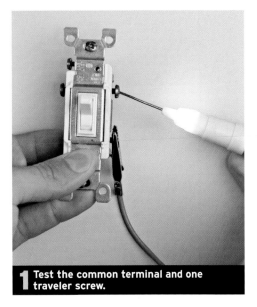

1 Test the common terminal and one traveler screw.

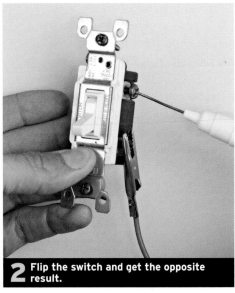

2 Flip the switch and get the opposite result.

TESTING A FOUR-WAY SWITCH

In addition to its green grounding screw, a four-way switch has four screws: two common (dark) screws at the top, and two traveler (brass) screws at the bottom ❹. To successfully test a four-way switch, move the test clip and probe to test all possible combinations, as shown by the arrows in the photos.

Start by testing both screws on each side of the switch ❺. Flipping the switch toggle should turn the switch off or on. After testing screw pairs on the same side of the switch, move the clip and probe diagonally. Again, flip the switch to turn the light off (if it was on) or on (if it was off).

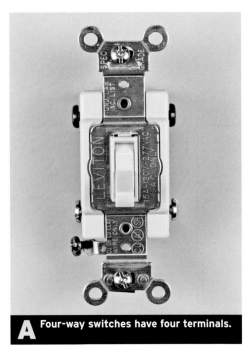

A Four-way switches have four terminals.

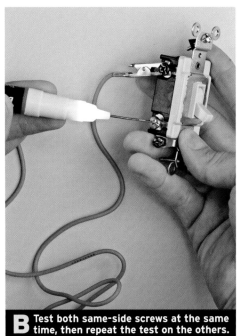

B Test both same-side screws at the same time, then repeat the test on the others.

WIRING A SINGLE-POLE SWITCH

The most commonly installed switch, a single-pole, is straightforward to wire. Spliced together during the rough-in stage, the neutral wires stay tucked in the outlet box. Pull ground and hot-wire groups out of the roughed-in box. Use the hole in the handle of your wire strippers or use needle-nose pliers to loop the conductor ends so they can be wrapped around the screw terminals.

First attach the ground wire to the green grounding screw on the switch **❶**. Orient the wire loop in a clockwise direction—the same direction the screw tightens. A loop facing the wrong way may be dislodged by the pressure of the screw head as it tightens down.

Next, connect the hot wires to the switch terminals, again orienting wire loops clockwise **❷**. One black wire is hot (power coming in), and the other is the switch leg (power going out to the fixture). With a single-pole switch, however, it doesn't matter which wire you attach to which screw. Generally, pros attach the hot wire last, much as they attach the hot wire on a receptacle last.

Once the ground and hot wires are connected to the device, they're ready to be tucked into the box **❸**. Always push the device into the box by hand until it's flush to the wall. Don't use screws to draw a device to a box because the device may not lie flat, and it's easy to strip the screw holes in a plastic box. Likewise, although using a screw gun is faster than a screwdriver, hand screw switches until you get the hang of it **❹**.

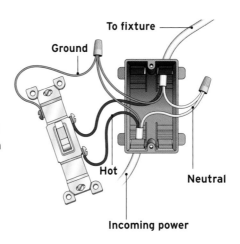

To fixture

Ground

Hot

Neutral

Incoming power

This switch controls a fixture at the end of a cable run. (See p. 174 for complete diagram).

1 Always attach the ground first.

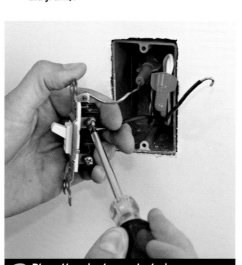

2 Place the wire loops clockwise on screws.

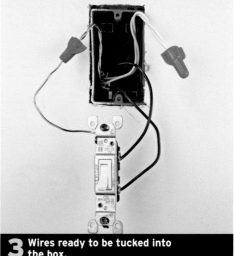

3 Wires ready to be tucked into the box.

4 Screw switches to the box by hand.

Wiring switches
Before connecting or disconnecting wires to a switch, use an inductance tester or a voltage tester to make sure that the power to the switch outlet is off. Test with the switch both on and off to be sure.

Because switches interrupt only hot wires, you'd think they'd all be easy to wire. As you'll see, however, switch wiring can also be quite complex, especially three-way and four-way switches and switches with electronic components. Follow the manufacturer's instructions carefully.

SINGLE-POLE SWITCH WITH BACK-FED WIRING

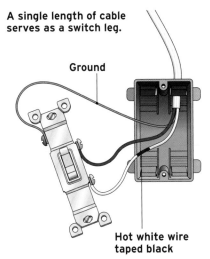

A single length of cable serves as a switch leg.

Ground

Hot white wire taped black

1 Tape the white wire black.

2 Connect the ground wire first.

3 Connect the switch leg.

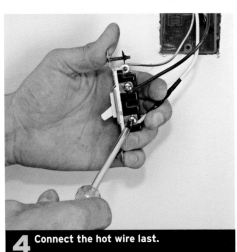

4 Connect the hot wire last.

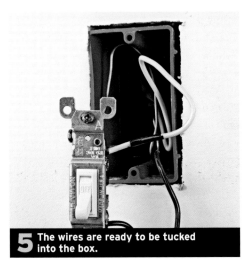

5 The wires are ready to be tucked into the box.

When an outlet box is closer to the power source than to the switch box, it's common to run a single length of 12/2 or 14/2 cable as a switch loop.

Splice the ground wires together and neutral wires together. If the fixture has a green ground screw, run a pigtail from the ground wires group and attach it to the screw. From the neutral wire group, run a pigtail to a fixture lead wire. Then splice the incoming hot wire to the two-wire cable running to the switch.

Note: Here, for convenience, we bend the rule of using a white wire only as a neutral wire and instead wind black tape on each end of the white wire to show that, in this case, the white wire is being used as a hot wire.

At the switch, start by stripping and looping the wire ends in the switch loop. Next, tape the white wire with black electrician's tape to indicate that it is serving as a hot wire to the back-fed switch. Convention dictates that the white wire in back-fed wiring is *always* the hot lead (power coming in) **1**. The black wire, on the other hand, is the switch leg that runs back to the fixture.

First, connect the ground wire to the green ground screw on the back-fed switch **2**. Next, connect the switch-leg wire (black) **3**,

then the hot wire (white taped black) to the switch terminals. To keep looped wire ends snug against the screw shaft as you tighten down the screw, pull gently on wires, as shown **4**. Not fumbling with wire ends saves time.

Finally, tuck the wires into the box, screw the switch to the box, and install the cover plate **5**.

REPLACING A SINGLE-POLE SWITCH

You can replace a single-pole toggle switch with a convertible dimmer that's wired as a single-pole dimmer. You can use the existing wires, but first turn off the power to the circuit. Use an inductance tester to see if voltage is present at the switch box **1** (most inductance testers can detect the presence of power even before you remove the cover plate). If the tester glows, there's power present: Turn it off at the fuse box or breaker panel. Test again. If the power's off, unscrew the old switch and pull it out from the box **2**.

Disconnect the switch wires and note their condition **3**. If the cable's fiber sheathing is frayed but individual wire insulation is intact, the wires are probably safe to attach to the replacement switch. If there's debris present in the box, sweep or vacuum it out.

Connect the wires to the new switch **4**. There may not be a ground wire to attach to the new switch's ground screw but Code doesn't require grounding a switch if there's

1 Test to make sure the power is off.

4 Wire the new switch to the existing wiring.

Cover all connections

All electrical connections must be housed inside a covered junction box so they can't be disturbed. Often, electricians will use an existing light box as a junction box in which to splice a cable feeding a new fixture. When there's not enough room in an existing box, use a separate junction box to house the splices.

Code requires all electrical connections to be housed in a junction box.

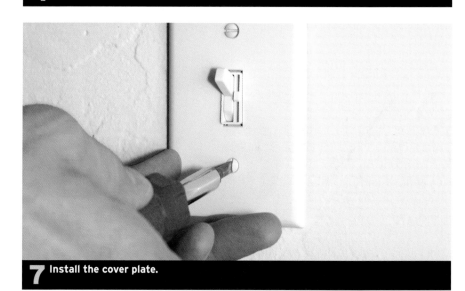

7 Install the cover plate.

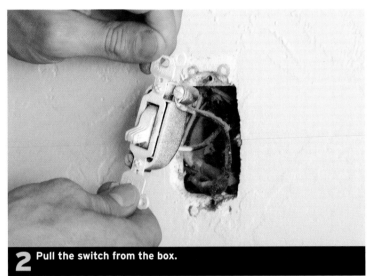

2 Pull the switch from the box.

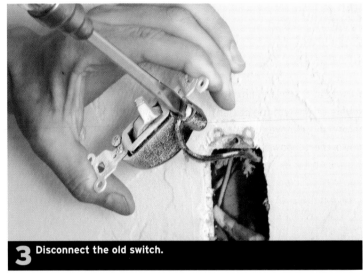

3 Disconnect the old switch.

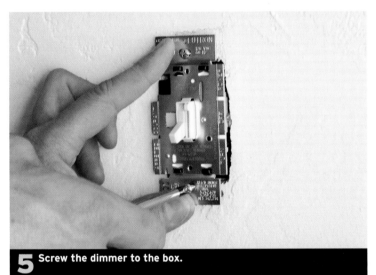

5 Screw the dimmer to the box.

6 The dimmer's rating is stamped on its face.

no ground wire feeding the box. Once the dimmer's connected, set it flush to the wall, and screw it to the box **5**.

Note: A dimmer must match the type of fixture it controls, whether incandescent or halogen or low-voltage. Typically, the dimmer rating is stamped on its face **6**. In this case, the rating specifies, "For permanent incandescent fixtures."

Finally, install the cover plate to protect the connections in the box and to prevent switch users from inadvertently touching the wire ends or dimmer terminals **7**.

TRADE SECRET
A convertible switch can be wired as a single-pole or a three-way switch. This type costs a bit more than a fixed switch but gives you more options on the job site. This switch may even save you a trip to the hardware store.

WIRING A LINEAR SLIDE DIMMER

Newer, more sophisticated dimmers often require different connectors—wire leads rather than screw terminals—but they're still basically switches and so interrupt only hot wires. For standard single-pole switches, it doesn't matter which screw terminals you connect a switch leg or hot wire to, but it does matter which wire you attach to dimmer leads.

Today's dimmers are sophisticated and expensive, so always read the directions that come with them. The slide dimmer shown in photo ❶ can be wired as a single-pole or three-way switch, depending on which wires you connect. Thus it has a bare-wire ground, red and black hot wires, and a yellow wire that is used to wire the dimmer as a three-way device ❷.

If the convertible device will be used as a single-pole dimmer, you won't need the yellow wire. So cap it with a wire connector ❸.

Splice the ground pigtail to the device's bare ground lead. Then splice the switch leg from the box to the red lead on the device. On devices with wire leads, typically a red lead attaches to the switch leg ❹.

Finally, attach the incoming hot wire to the other hot lead (black) on the device ❺. Carefully fold the wires into the box and push the wired dimmer into the box. Screw the device to the box, and install the cover plate.

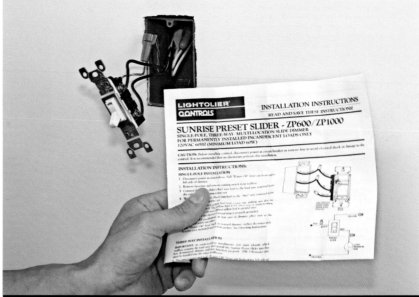

1 Dimmer specifics can differ from one manufacturer to the next. Read the directions before installation.

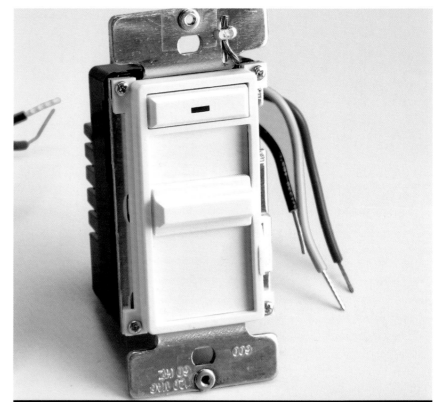

2 Prepare the leads before you begin attaching wires.

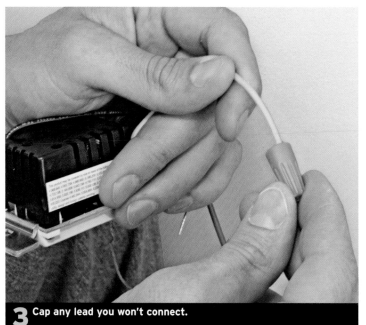

3 Cap any lead you won't connect.

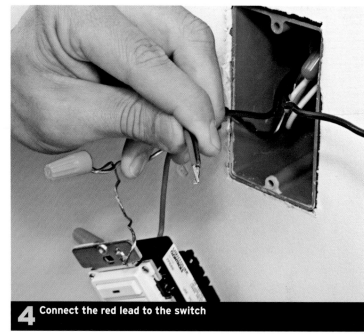

4 Connect the red lead to the switch

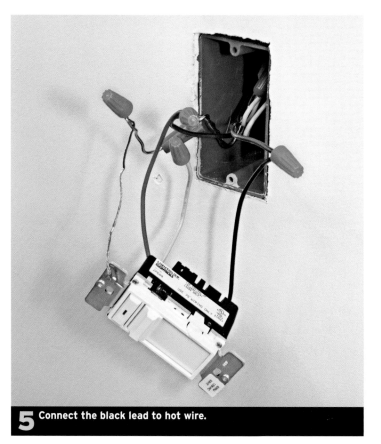

5 Connect the black lead to hot wire.

Dimmer leads It's critical which circuit wires you attach to dimmer leads. So during the rough-in stage, distinguish which wire is the switch leg and which is the incoming hot wire for each single-pole dimmer that you install. To distinguish these wires, many electricians wrap the wires together and bend back the switch leg wire. Typically, the red dimmer lead attaches to the switch leg wire, and the black dimmer lead attaches to the incoming hot wire. Or slip a short sleeve of cable sheathing back onto the stripped wires and use a felt-tipped marker to label what each wire does.

ROUGHING IN A THREE-WAY SWITCH

Three-way switches allow you to operate a light from two locations. They're often used at the top and bottom of a set of stairs or at two entrances to a room. If you get confused about which wire goes where, refer to the wiring schematics on pp. 174–176 or make a drawing of your own.

In new wiring, wires are roughed in when the framing is still exposed. Here, 12/2 and 12/3 cables were fished in to feed a three-way switch that was added after the drywall was up.

➜ **For more on fishing cables, see p. 193.**

After stripping sheathing from the 12/2 cable, strip the 12/3 cable. Removing 12/3 sheathing is a little different: Start by lightly *scoring* the 12/3 cable along its length, up into the box **❶**. Then, when you reach the end of the cable, *cut* through the sheathing. Because you'll soon be stripping the ends of individual wires, cutting through the sheathing end won't compromise wire insulation.

➜ **For more on stripping cable, see p. 37.**

Starting at the cut-through sheathing at the end of cable, pull the sheathing free of the wires within **❷**. The sheathing will separate easily along the scored line. At the cable's upper end in the box, carefully cut free the sheathing.

Twist together the ground wires before splicing them with a wire connector **❸**. To use a green wire connector, cut one of the leads shorter than the other so that it sticks out of the hole in the end of the wire connector. Then connect that ground lead to the switch's green ground screw. Next, strip wire insulation from the neutrals, splice them, cap them with a wire connector, and push them into the box **❹**. (Neutrals don't connect to standard switches.)

After professional electricians strip cable sheathing, many wrap individual wires in a distinct fashion so any other electrician will know which wires are travelers and which are switch-leg wires. Group and twist traveler wires clockwise, then wrap the switch-leg wire counterclockwise about the travelers. This way there's no need to put tape or labels on the wires to identify them or to pull out all the cables and figure out which wire is what **❺**.

1 Score the 12/3 cable along its length.

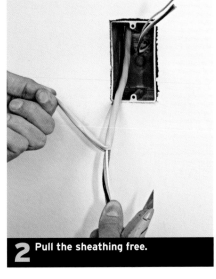
2 Pull the sheathing free.

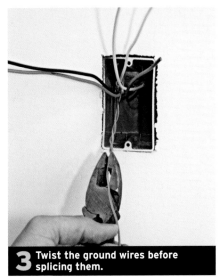
3 Twist the ground wires before splicing them.

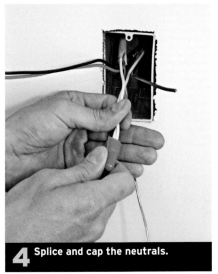
4 Splice and cap the neutrals.

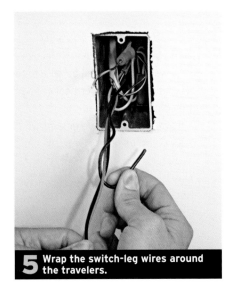
5 Wrap the switch-leg wires around the travelers.

WIRING A THREE-WAY SWITCH

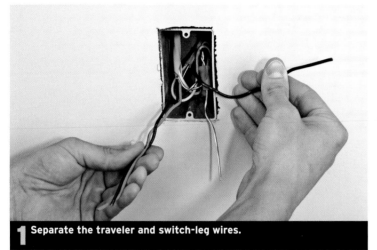

1 Separate the traveler and switch-leg wires.

2 Strip the ends of the wire.

3 Attach the ground, then a traveler.

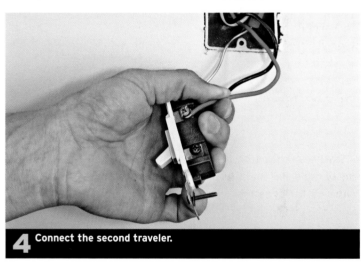

4 Connect the second traveler.

To wire a three-way switch, pull the conductors out of the outlet box. Unwrap the switch leg wire from around the travelers and separate the wires so the travelers are on one side and the switch leg on the other ❶. If individual wire ends weren't stripped during the rough-in phrase, do so now. Give the strippers a quick twist, use your thumb as a fulcrum to push the wire insulation off, and loop the wire ends ❷.

After attaching the bare ground wire to the green grounding screw on the device, attach the first traveler wire ❸. Loop wires clockwise around the screw shafts, and they'll be less likely to slip off when the screws are tightened. Flip the switch over and connect the second (red) traveler ❹. (*Note:* It doesn't matter which traveler wire goes where—you'll still be able to turn lights off and on. The only critical connection is the common terminal.)

If you connect traveler wires in the same position on two three-way switches—say, you attach the red traveler to the first terminal, as just described—the lights will be off when switch toggles are both up or both down. This is a fine point and most people needn't agonize about it: The three-way switches will still work as long as you attach traveler wires to traveler terminals.

After attaching the second traveler, connect the hot conductor (switch leg or hot wire) to the common screw terminal, which is color coded black ❺. Push the device into the box by hand, screw the device to the box, and install a cover plate.

Photo ❻ is a frontal view of the switch we just wired; it's typical for a three-way that's located between the power source and the light fixture it controls.

→ Also see "Wiring a Back-fed Three Way Switch," p. 60.

>> >> >>

WIRING A THREE-WAY SWITCH (CONTINUED)

WIRING A BACK-FED THREE-WAY SWITCH

For a back-fed three-way switch, a single three-wire cable feeds the switch located beyond the light fixture. Hot wires run from a splice in the fixture box. In this case, the red and black wires are travelers. The white wire here is not a neutral; thus it is taped black to indicate that it is the hot conductor connected to the common terminal.

→ **For more on wiring back-fed switches, see the drawing on p. 53.**

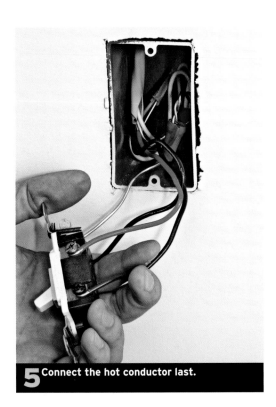

5 Connect the hot conductor last.

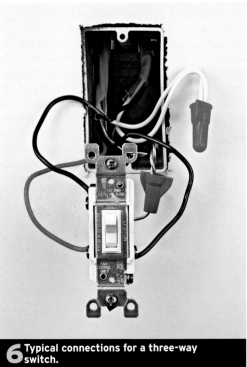

6 Typical connections for a three-way switch.

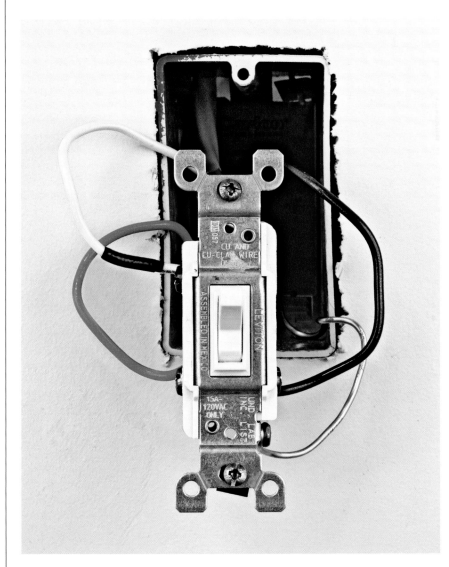

This is a typical **three-way switch with back-fed wiring.**

WIRING A FOUR-WAY SWITCH

Four-way switches have two travelers incoming (from the power source) and two travelers outgoing (to a second four-way switch). Thus there will usually be two three-wire cables entering the box. (In the sequence shown here, you know we're working with 14/3 cable because the cable sheathing, visible in the box, is white.)

Twist and splice the neutral wires with wire connectors and push them out of the way, into the back of the box. Then strip and loop the ends of the ground wire and the hot conductors and twist them so they turn clockwise around the screw shaft. Next, connect the ground to the four-way switch, as you do for all devices ❶.

Connect a set of travelers on one side the switch. On each side of a four-way switch, there is a dark terminal and a brass terminal: Connect similar wires to similar terminals. You can connect red wires to dark terminals or black wires to dark terminals—it doesn't matter—just be consistent on both sides of the box. Note how the electrician exerts a slight tension on the wires to keep them from slipping off as he tightens the screw terminals ❷.

To summarize how the wires connect to the four-way switch in this photo sequence: Red travelers are attached to brass screws; black travelers, to dark screws ❸. When you've got this many wires in a box, it's helpful if you partially accordion fold the wires before you push them into the box. Screw the switch to the outlet box, then install the cover plate.

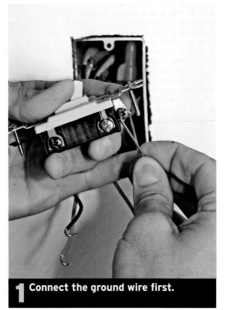

1 Connect the ground wire first.

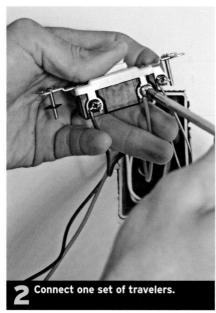

2 Connect one set of travelers.

3 Similar wires connect to similar terminals.

> **TRADE SECRET**
> When replacing a defective or outdated three- or four-way switch, use a felt-tipped marker to note which wires connect to which switch terminal *before* disconnecting the wires. Of course, turn off power to the switch before you begin.

WIRING A SWITCH/RECEPTACLE COMBO

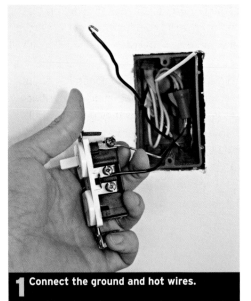

1 Connect the ground and hot wires.

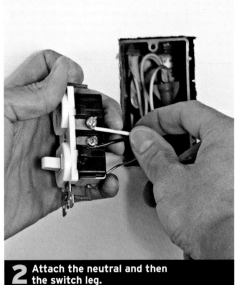

2 Attach the neutral and then the switch leg.

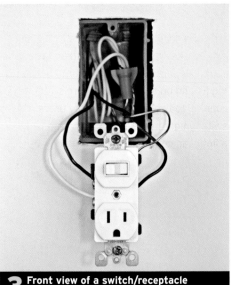

3 Front view of a switch/receptacle combo.

Switch/receptacle combinations are often installed along kitchen counters, where space is at a premium, because it enables you to fit a switch and a receptacle into a single outlet box. In effect, you'll be wiring both a single-pole switch—which interrupts hot wires—and a receptacle that will always be hot. The switch requires a hot wire (from the power source) coming in and a switch leg going out, and the receptacle requires a hot and a neutral wire. A ground wire connects to the device, too.

➡ **Specialty switches are shown in the bottom photo on p. 40.**

In the sequence shown here, two 14/2 or 12/2 cables feed the setup. During the rough-in stage, splice ground and neutral wires and run a pigtail from each group, which will be connected to the device later.

Connect the ground pigtail to the green ground screw on the device; then connect the hot leg **1**. *Note:* The tab between the two brass screws has not been broken out, so the hot wire will feed both the switch and the receptacle.

Turn over the switch. Notice that, on this side of the device, there is a brass screw and a silver screw. (The two screws are physically isolated from each other inside the device.) Attach the neutral (white wire) to the silver screw, which serves the receptacle. Then connect the black switch leg to the brass screw nearest the switch **2**.

The wired device as seen from the front is shown in photo **3**. On this device, the two screw terminals of the single-pole switch are on opposite sides.

WIRING A PILOT LIGHT

Pilot light switches are used when the light fixture is on the other side of a door that is usually closed, such as a basement door or the door to a walk-in cooler. This specialty switch requires a neutral wire so its tiny pilot light will glow. Wiring this device is similar to wiring a switch/receptacle combo: A single hot wire feeds both the switch and the tiny light bulb, a neutral serves the light, and a black switch leg runs from the switch to the light fixture in the next room.

The pilot lights let you know whether the light on the other side of the door is on.

WIRING A DOUBLE SWITCH

This economy switch has only push-in (back-wired) terminals and no ground screw. On one side is a single hot lead coming in, and on the other side are switch legs running to two different energy users. This switch might be used in a bathroom with a light/fan combination; one switch would control the fan and the other, the light. This setup would require one 12/2 cable coming in to supply power and a 12/3 cable going out.

Note: Putting a bath fan on a timer is preferable to wiring it with an on/off switch. A timer allows the fan to continue running a while after the user has left the bathroom and turned off the light.

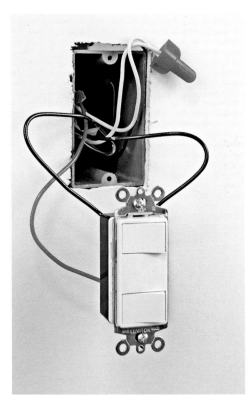

A stacked or double switch **controls power to two energy users.**

WIRING A MOTION DETECTOR

Motion detectors are specialized switches that sense motion by infrared or ultrasonic sensors and then turn on a light. The unit will remain on as long as there is sound or motion present and for a fixed interval thereafter. Then it will shut itself off automatically.

Because manufacturing details vary widely, it's difficult to generalize about wiring motion detectors; the switch's location in the layout—midcircuit or back fed—will also dictate the type of cables feeding it and how they're connected to switch leads. Some motion detectors require a neutral connection; others don't. The motion detector shown here has no neutral connection. Its green wire is a ground lead, the back lead connects to the incoming hot wire, and the blue wire connects to the switch leg **❶**.

This motion detector's sensor is the large "eye" at the top of the unit. Finally, secure the cover plate **❷**.

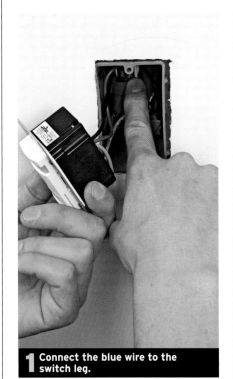

1 Connect the blue wire to the switch leg.

2 Secure the cover plate to protect the components.

⚠ WARNING

Specialty switches often combine several functions and so may require more complex hookups. Reading the directions will ensure a correct installation and prevent you from damaging the switch. Switches with solid-state electronic components are particularly sensitive.

WIRING AN ELECTRONIC TIMER SWITCH

An electronic timer switch is a sophisticated piece of equipment. In addition to a control screen and several programming buttons, the device is surrounded by a metal fin called a *heat sink*, which dissipates the heat generated by resistance within the switch, thus prolonging the life of the device. Better quality dimmers and programmable switches tend to have heat sinks.

This timer switch has four leads, including a neutral, so two 12/2 cables feed this box. Start by splicing the stranded green ground lead directly to the incoming ground wire group—no need to run a separate ground pigtail to the switch **❶**.

Unlike most mechanical switches, this timer switch has electronic components that require a neutral wire to operate—hence the timer's neutral (white) lead is spliced to a neutral wire group from the cable feeds. Splice the hot (black) wire to the black lead of the switch and the switch leg to the blue wire of the switch **❷**. Again, read the directions to be sure you're installing it right.

After carefully folding the spliced wires into the box, hold the face of the switch flush to the wall and screw it into the box **❸**. Then install the cover plate **❹**. (*Note:* The timer switch's control screen, large override button, and three smaller programming buttons are at the bottom.)

TRADE SECRET
When splicing a stranded wire to solid conductors, as shown in photo 1, place the tip of the stranded wire slightly beyond the solid conductors so that when you twist on the wire connector, its threads will fully engage the stranded wire and ensure a solid splice.

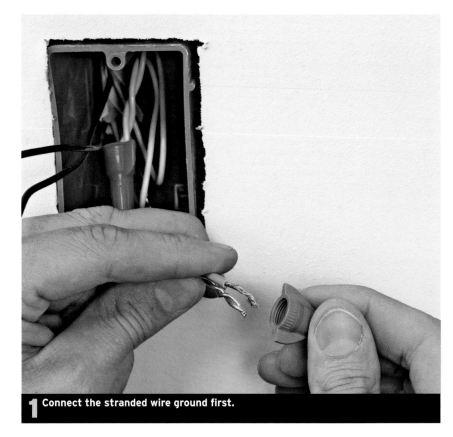

1 Connect the stranded wire ground first.

3 Hold the timer face flush.

WIRING A MANUAL TIMER

Manual timers tend to be inexpensive and simple to wire: They're basically a single-pole switch. The model shown here has no ground lead, and its lead wires attach to a hot wire and a switch leg. Because both leads are black, it probably doesn't matter which lead you connect to the hot wire or switch leg. After installing the cover plate, snap the plastic dial onto the metal post in the middle of the unit.

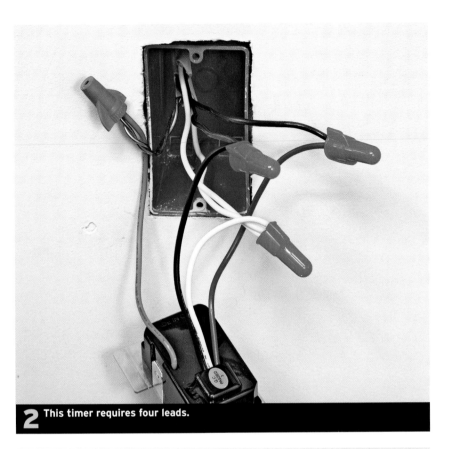

2 This timer requires four leads.

A manual timer **turns off a light after a set period of time.**

4 Install the cover plate.

LIGHTING

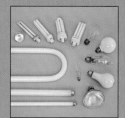

BASICS

REPAIRS

BASIC FIXTURES

WALL SCONCES

CHANDELIERS

RECESSED LIGHTING

UNDER-CABINET FLUORES-CENTS

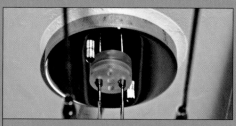

LOW-VOLTAGE FIXTURES

MONORAIL TRACK LIGHTING

FIXTURE WIRING

The temporary lighting socket in the photo below shows the basics of connecting a light fixture to *supply or source wires* The supply ground wire connects to a green fixture pigtail, its neutral (white) supply wire connects to the neutral pigtail, and its hot (black) supply wire connects to the hot pigtail.

Connect wires in this order: ground, neutral, and then hot. And disconnect them in the reverse order: hot, neutral, and then ground. In this manner, the grounding system stays connected as long as possible to protect you. Remember to ground metal boxes, mounting plates, and fixtures.

The basics of light fixture wiring. (The bare copper and green wires are partially covered with white paint.)

Wiring, Lighting, and Low-Voltage Terms

Hickey
A threaded coupling that joins two lengths of threaded tubing.

Lamp
For most of us, a light fixture that sits on a table or floor. To an electrician, *lamp* is the preferred term for a light bulb: A standard incandescent bulb is thus a "type A lamp." To avoid confusion in this book, a lamp is a fixture, not a bulb.

Leads
Wires preattached to a fixture. Leads are spliced to supply wires to energize the fixture. Typically, a fixture has ground, neutral, and hot leads. Fixture leads are often stranded wire.

Light Box
An outlet box that serves a light fixture.

Line Voltage
The standard current in most house circuits: 120v.

Lumens
A measure of light on the surface of a bulb.

Nipple
Short section of hollow threaded rod.

Lo Vo or LV
Low voltage. These systems are typically 12v but sometimes 24v.

Omnidirectional Flare
Light from a standard incandescent bulb that radiates in all directions; as opposed to a controlled or directional beam.

Primary Wires
In a lo-vo system, wires running from a 120v power source to the transformer.

Running Thread
Hollow threaded rod or pipe in the center of a light fixture, which provides a conduit for wires and a way to connect various fixture parts.

Secondary Wires
Wires running from a transformer to a lo-vo tracks, cables, or fixtures.

Source or Supply Wires
Wires from the power source, typically 120v.

Transformer
An electrical device that reduces line voltage to low voltage. All lo-vo lighting systems require a transformer.

BULB TYPES

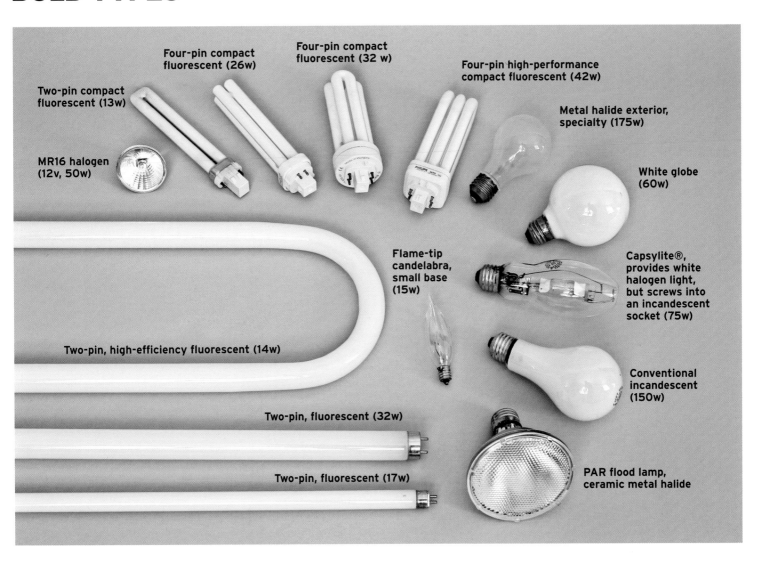

Two-pin compact fluorescent (13w)

Four-pin compact fluorescent (26w)

Four-pin compact fluorescent (32 w)

Four-pin high-performance compact fluorescent (42w)

Metal halide exterior, specialty (175w)

MR16 halogen (12v, 50w)

White globe (60w)

Capsylite®, provides white halogen light, but screws into an incandescent socket (75w)

Flame-tip candelabra, small base (15w)

Two-pin, high-efficiency fluorescent (14w)

Conventional incandescent (150w)

Two-pin, fluorescent (32w)

Two-pin, fluorescent (17w)

PAR flood lamp, ceramic metal halide

There are more bulb types than there's room to describe them, so we'll stick to the three most common types here: incandescent, fluorescent, and halogen. When changing burned-out bulbs, make sure that the replacement bulb fits the fixture socket because, even within each bulb type, there are variations in diameter, pin size (for fluorescents and halogens), and so on. You should *never* have to force a bulb to make it fit a socket.

Incandescent bulbs

Incandescent bulbs screw into threaded sockets and complete a circuit by making contact with the silver shell (neutral) and the contact tab (hot) in the bottom of the socket. Typically, the socket is energized with house current (120v). Incandescent bulb wattage is stamped on the top of the bulb.

Fluorescent bulbs

Fluorescent bulbs have contact pins at either end that slide into fixture sockets and then rotate into final position; two-pin bulbs plug into magnetic ballasts, four-pin bulbs plug into electronic ballasts. A fluorescent tube increases wattage as it increases in length, it can also coil to increase wattage. Fluorescent bulbs are more energy efficient than incandescent ones, so many building codes now mandate that a certain percentage of house lighting be fluorescent.

Halogen bulbs

Halogen bulbs can be installed in both standard and low-voltage (lo-vo) systems. But even on reduced current, they burn bright–and hot. The filament in a halogen bulb reaches in excess of 1200ºF, so be sure to install these bulbs where they can't be accidentally touched.

TESTING LAMPS & REMOVING OLD WIRES

If a lamp flickers or doesn't light at all, you can perform a series of tests to identify the problem. A chandelier is more complicated because it will have several sockets, but investigating its problems is essentially the same. Perform these tests with the lamp unplugged or the chandelier disconnected.

Scrub the inside of an old socket with steel wool to remove corrosion and improve electrical contacts ❶. You might also use needle-nose pliers or a screwdriver point to lift the contact tab in the middle of the socket. Screw in a light bulb and see if that repair helps. If it doesn't, remove the socket.

Press the sides of the socket together to pop it out of the base. Disconnect the wires attached to the socket and test it. Attach the continuity tester clip to the socket's brass screw terminal (hot) and the tester probe to the contact tab in the bottom of the socket ❷. Turn the switch off and on. If the tester doesn't light, the socket is defective and should be replaced.

If the socket is okay, test the cord. Connect the tester clip to the narrow plug prong and the tester point to the bare wire ends that were connected to the socket—or still are. Touch the tester point to both wires, move the clip to the wide prong, and then repeat the test ❸. If there's no continuity in either (or both) of the wires, there's a break in the wire or faulty connections to the plug. Replace the cord and plug.

Finally, if the lamp base is metal, test for shorts by attaching the tester clip to the narrow (hot) prong of the plug, and touching the tester point to the lamp metal base. Switch the lamp off and on. If there's a short, the tester will light ❹.

If lamp parts are defective or unsightly, squeeze the socket shell to remove it from the socket base: Typically, "press here" is stamped on the socket shell. Slide off both the shell and the cardboard liner ❺. If you're replacing only the switch, unscrew the wires from its terminals ❻.

However, if you're replacing the cord and plug as well, simply snip one end of the lamp cord and pull the cord out of the lamp.

To replace entire socket assembly, unscrew the set screw that holds the socket base to the threaded rod that runs through the center of the lamp. Then turn the socket base counterclockwise to remove it ❼.

1 Use steel wool to clean and improve electrical contacts.

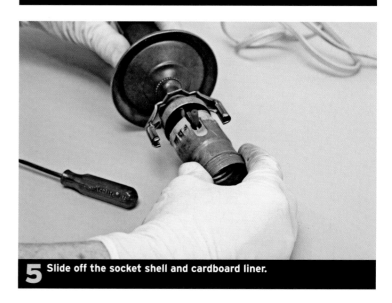

5 Slide off the socket shell and cardboard liner.

TRADE SECRET
Take photos of lamps and fixtures before you disassemble them, and the fixtures will be easier to put back together. To avoid losing hard-to-find and tiny parts, put similar pieces in plastic bags and label them.

2 Test the brass screw and socket tab.

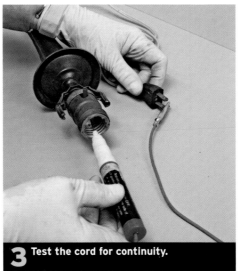

3 Test the cord for continuity.

4 Test metal lamps for shorts.

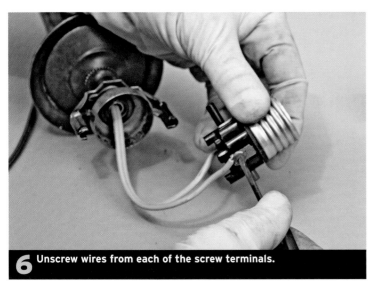

6 Unscrew wires from each of the screw terminals.

7 Unscrew the socket base from the threaded rod.

Work safely Unplug lamps or shut off power to chandeliers before working on them. If a chandelier is presently connected to a circuit, use a voltage tester to make sure the power is off before disconnecting the fixture. If a fixture is heavy, by all means get help supporting it and lowering it after it has been safely disconnected.

Wear work gloves to protect your hands; wires, sockets, and other metal parts are often sharp. Many repair pros favor disposable rubber or latex gloves, which are flexible enough for fine work, but impervious to cleaning solvents and toxic metal dusts. If you use power tools to drill, grind, or polish metal or if you solder connections, wear a respirator mask and eye protection.

REWIRING A LAMP

1 Screw on the new socket base.

2 Pull the clipped end of the replacement cord through the lamp.

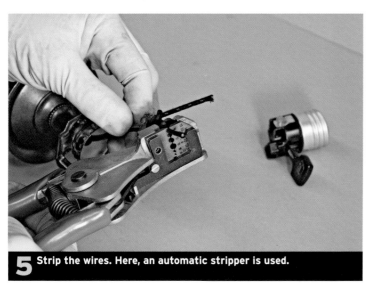

5 Strip the wires. Here, an automatic stripper is used.

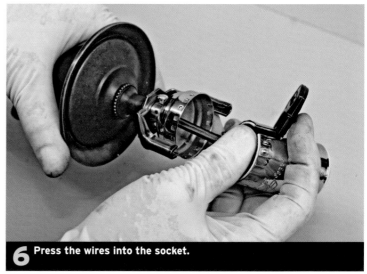

6 Press the wires into the socket.

After the old wires have been removed, a fixture is ready for rewiring. Screw the new socket base onto the threaded rod until the base is snug against the curved harp saddle (the harp, when fitted into its saddle, supports the lamp shade) **1**. Cut the replacement cord to length. Typically, that's about 9 ft.: 6 ft. of clear cord from the lamp to the plug, 2 ft. for the cord hidden inside the lamp, and roughly 1 ft. above the socket, which you'll strip, tie, and trim **2**.

If the cord has a fabric sheathing, use a utility knife to slit it along its length. Then pull apart the two wires inside the cord **3**. Next, tie an underwriter's knot, which prevents the cord from being pulled back into the lamp and stressing connections to switch terminals **4**. Trim the wire so there's about 1½ in. above the knot. Then strip the wire sheathing to expose bare wire. Although you can use a standard wire stripper, the automatic wire stripper shown here severs and pops off the sheathing with one squeeze of the tool **5**.

Twist and loop the bare wire ends . Then place the loops on the screw terminals of the switch—make sure they loop clockwise. Tighten the screws down, using your thumb to press the wires into the bottom of the socket **6**. Slide the cardboard liner over the wired switch. Then slide the socket shell over the liner. Rock the shell into its base until you hear two clicks. *Note:* Attach the marked or ridged wire to the silver (neutral) screw on the switch **7**. After snapping a socket shell into its base, gently wiggle the shell to make sure it's seated solidly.

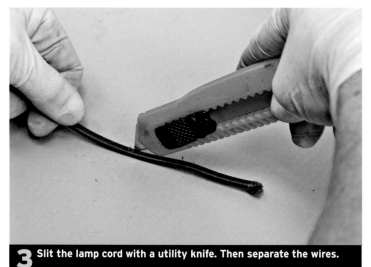

3 Slit the lamp cord with a utility knife. Then separate the wires.

4 Tie an underwriter's knot to secure the cord in the lamp.

7 With completely new wiring, this old lamp is as good as new.

Tinning Tinning, often referred to as soldering, stranded wires is not absolutely necessary, but pros do it because it fills in spaces between wire strands. In effect, soldering makes stranded wire solid, so it's unlikely to crush and spread out when a screw head tightens on it. Tinned strands are also easier to twist wire connectors onto. *Note:* Use only rosin-core solder for electrical connections; acid-core solder will corrode them. And make sure you solder only on a heat-resistant surface.

Soldering makes stranded wire **solid and easier to work.**

REPLACING A PLUG

1 The points of a quick plug pierce the lamp cord.

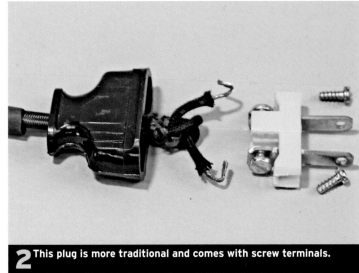

2 This plug is more traditional and comes with screw terminals.

You should replace any plug that is cracked, difficult to remove, or whose cord is damaged near the plug. All lamp cord is 18 gauge and will fit any type of lamp plug. If the plug's prongs are different sizes—one prong is wider—then the plug is polarized.

Individual wires in lamp cord are also differentiated so you can attach them correctly to switch and plug terminals.

If the lamp cord is plastic sheathed, the wire to be connected to neutral terminals will be ridged; the wire to be connected to hot terminals is smooth. If the cord is sheathed with fabric, remove the sheathing and you'll see that one wire is striped or marked in some way—that's the neutral.

There are two main types of lamp cord plugs. Connect plastic-sheathed lamp cords to a "quick plug" **❶**. Cut the end of the cord square, thread it through the plug shell, and insert it into the body of the plug. As you squeeze the plug prongs together, two sharp points on the prongs pierce the sheathing to create an electrical connection. Slide the plug shell over the body to lock the cord and prongs in place.

If you have fabric-sheathed cord, however, use a plug with screw connections. Strip about 1 in. of fabric sheathing, then strip

about ½ in. off each wire end to expose bare wire. Thread the wire through the plug body, tie an underwriter's knot, loop each wire clockwise around a screw terminal, then tighten the screws **❷**.

Polarized fixtures

Light fixtures parts are polarized, so they fit together only one way. On a correctly wired light fixture, the neutral wire of the lamp cord connects to the silver screw on the socket, whereas the hot wire connects to the brass screw on the socket.

If the hidden parts of the socket were visible, you'd see that the silver screw makes contact with the sides of the socket shell (the part that the bulb screws into), and the brass screw makes contact with the tab in the bottom of the socket. Should you inadvertently touch the side of the shell while

changing a bulb, there's no harm done if the socket is correctly wired. However, if you reverse the order in which wires are attached—known as reversing the polarity—the outside of the screw shell becomes hot (energized) and could shock someone changing a bulb.

To ensure that you connect light fixture wires correctly, lamp cord wires are differentiated. Inside fabric-covered cords, the insulation of the wires will have different colors or the neutral wire insulation will be stripped or marked in some distinctive manner. If the lamp cord is plastic, the neutral wire on the cord will have ribs or ridges; the hot wire will be smooth.

Finally, the neutral wire of lamp cord connects to the wider prong of a plug; the hot wire connects to the narrow prong of a plug.

TRADE SECRET
Near the plug, slide a piece of heat-shrink tubing over the end of a fabric-covered lamp cord to keep it from unraveling.

REFURBISHING A CHANDELIER

Chandeliers vary wildly, especially those designed in the early decades of the 20th century. Fortunately, you can get replacement parts for many chandeliers and substitute modern equivalents for many other parts. Still, it's wise to save all original parts of a fixture. Photograph the fixture before you take it apart, and you'll know what goes where when it's time to reassemble it.

Survey the chandelier for missing or damaged parts. As you disassemble the unit, put like parts in plastic bags and label them. In many cases, old switch and socket assemblies will still work. Use a continuity tester to test old switches, and lightly rub steel wool inside sockets to remove corrosion and improve electrical contacts.

Old wiring is rarely worth saving, however. At best, it will be brittle and, at worst, broken or unsafe. Snip the wires running to each socket ❶ (see p. 76) so you can remove the sockets from the shade holders. Removing sockets takes finesse and patience. Socket shells snap into socket bases, so rock the shells gently from side to side until they unsnap. Then angle the shells out of the holder ❷ (see p. 76). Chances are, the cardboard liners (sleeves) inside the shells will have deteriorated and should be discarded.

In many older chandeliers, source and switch wires are spliced together and housed in the fixture body. To disconnect the body, unscrew the finial at the bottom. Separate the cover (top) and pan (bottom) of the body, disconnect the wire splices and remove the old wires. Be sure to save the nipples (short threaded rods) and hickey (U-shaped coupling), because you'll need them to reassemble the chandelier body.

Rewiring the fixture

Before rewiring a chandelier, refinish it or simply clean up its parts. These days, leaving the surface patina intact is in vogue; the fixture seen here was washed with a simple household cleaner, rubbed with steel wool to

>> >> >>

WIRING A VICTORIAN LIGHT FIXTURE

Connecting incoming wires to fixture leads is standard: hot to hot, neutral to neutral. Grounds are spliced and connected to metal boxes and fixture ground screws, if any. *Note:* Fixture bodies and mounting devices vary considerably.

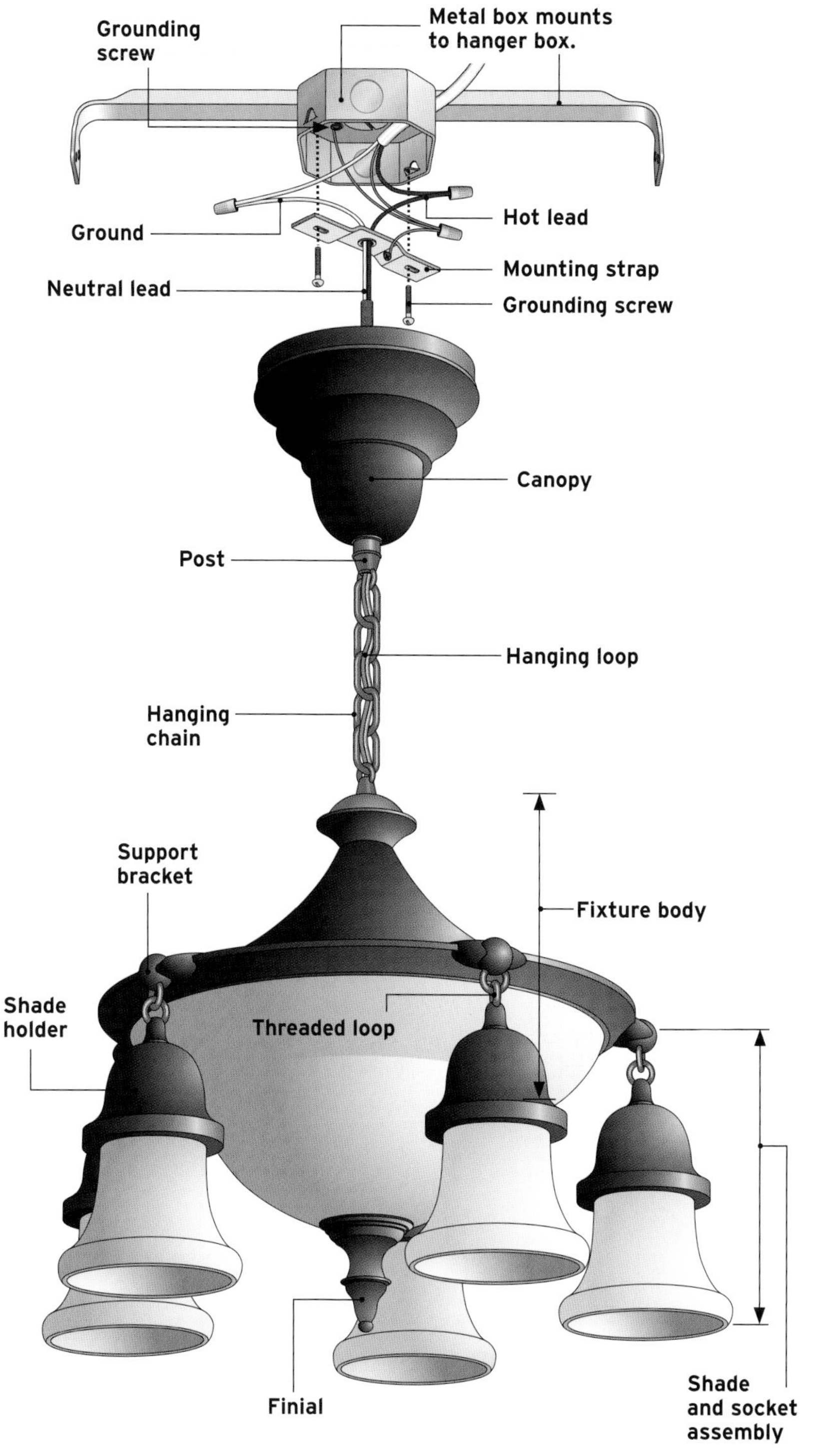

- Grounding screw
- Metal box mounts to hanger box.
- Ground
- Hot lead
- Neutral lead
- Mounting strap
- Grounding screw
- Canopy
- Post
- Hanging loop
- Hanging chain
- Support bracket
- Fixture body
- Shade holder
- Threaded loop
- Finial
- Shade and socket assembly

REFURBISHING A CHANDELIER (CONTINUED)

remove rusted spots and other obvious flaws, and then lacquered. Because the fixture's sockets still worked after 80 years, the restorer decided to reinstall them and wire them with rayon lamp cord. Use a utility knife to remove about 1½ in. of the fabric sheathing, and then use wire strippers to strip ½ in. of insulation off the individual wire ends. Loop the wire ends and place them clockwise onto the socket screws—in the same direction that the screws tighten ❸. For most lamp cord, individual wires are different colors; in this example, the installer made the blue wire neutral and the brown wire hot. It doesn't matter what color you designate hot or neutral, as long as your designations are consistent.

After wiring each socket, slide on a new cardboard liner to insulate the screw terminals. The socket base should also have a new liner ❹. Snap each socket shell to a base. Then use a small screwdriver to tighten the threaded part of the socket base to a male threaded loop linked to the fixture body ❺. (A screwdriver won't mar the fixture surface as pliers would.)

Feed the wires from each socket into the fixture body ❻. Group the neutral and hot wires from all the sockets. Then splice each group to the main lead wires, which you'll eventually connect to the source wires (neutral to neutral, hot to hot). Although it's not essential, professionals prefer to tin (solder) the ends of stranded wires, using rosin-core solder, so the wires'll splice better. Use wire connectors to splice each group, tug gently to be sure wires are solidly spliced, then wrap each splice generously with electrician's tape ❼.

Reassemble the fixture

Once you've spliced socket wires to main lead wires, feed those leads through the opening in the hickey and into the threaded rod that runs through the top of the fixture body ❽. Then stack the chandelier pan, cover, and related parts onto the hickey-and-threaded-rod assembly ❾.

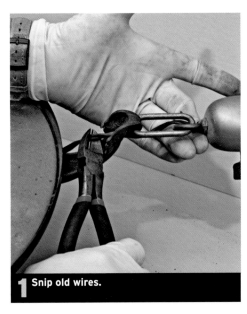

1 Snip old wires.

2 Carefully rock the sockets out of their holders.

6 Feed the socket wires into the body of the fixture.

7 Wrap each of the splices with electrical tape.

Chandelier assemblies vary, so be guided by the photograph you took of your fixture before disassembling it. Typically, a finial screws to the bottom nipple, and a female threaded loop to the top nipple. After the main leads emerge from the top nipple and loop, thread the leads through the hanging chain that support the chandelier body. Then feed the leads into the hollow post that the canopy screws to. Leave roughly 1 ft. of

free cord above the post, to strip and attach to source wires. (The power will be off, of course.)

Chandeliers' mounting details also vary, so, again, be guided by your fixture. If your old fixture has a crow's foot, remove it, because it was intended to screw directly to a ceiling joist and doesn't allow much flexibility. Replace it with a standard mounting bar, which screws to a properly rated

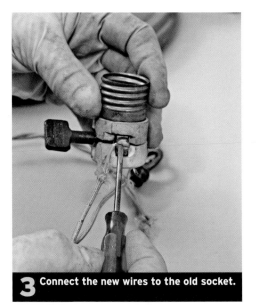

3 Connect the new wires to the old socket.

4 Install the liners on the socket and base.

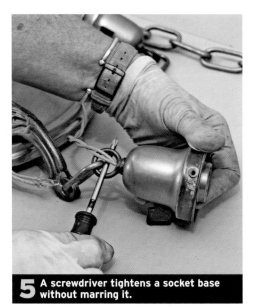

5 A screwdriver tightens a socket base without marring it.

8 Feed main leads into the hickey.

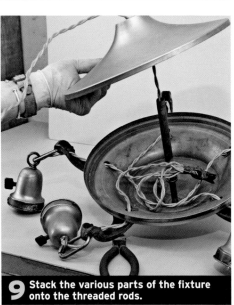

9 Stack the various parts of the fixture onto the threaded rods.

10 Remove the old crow's foot mount, replace with new mounting bar.

ceiling box. The threaded post under the canopy screws into a threaded hole in the center of the mounting bar **10**.

Shade holders are usually held in place with three small thumbscrews; tighten them just snug and then back off the screws a fraction so the shades will have room to expand without cracking when they heat up **11**.

11 The refurbished chandelier is as good as new but has an antique charm.

MOUNTING LIGHT FIXTURES

CEILING FIXTURE ELEMENTS

In this basic setup, the ceiling box mounts to an adjustable bar, which is screwed to ceiling joists. The fixture, in turn, screws to a mounting bracket, which screws to the ceiling box. All metal boxes and brackets must be grounded to be safe. Many electricians use grounding screws in both the box and the bracket, but one ground is sufficient: The metal mounting screws provide grounding continuity to box and bracket.

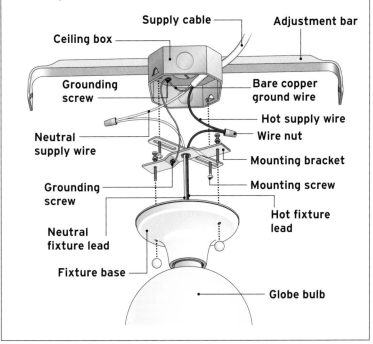

Supply cable
Adjustment bar
Ceiling box
Grounding screw
Bare copper ground wire
Hot supply wire
Wire nut
Neutral supply wire
Mounting bracket
Mounting screw
Grounding screw
Hot fixture lead
Neutral fixture lead
Fixture base
Globe bulb

Fixture boxes must be mounted to framing to adequately support the weight. (Boxes that are not solidly mounted are a hazard because in time they'll move, and that movement could compromise electrical connections inside.) There are many mounting options for boxes: The main choice is whether you nail or screw the box directly to a stud or ceiling joist or use an extendable mounting bar to which the box attaches, as shown here.

Either method works fine, but because the box slides along a mounting bar, you position the box—and hence the light fixture where you want. The other choice for a recessed can, which is also adjustable, is a housed unit such as that shown in the bottom photo on the facing page.

Mounting fixtures to boxes

If mounting screws on all light fixtures were exactly the same diameter and spacing as the screw holes on all boxes, life would be simple and you'd screw the fixture directly to the box. But there are many different box sizes and configurations, and light fixtures vary considerably. Consequently, there are many mounting brackets to reconcile these differences. Always examine existing outlet boxes before buying new fixtures and make sure that fixture hardware can mount to existing boxes. Otherwise, a routine installation could turn into a long, drawn-out affair with a lot of trips to the hardware store.

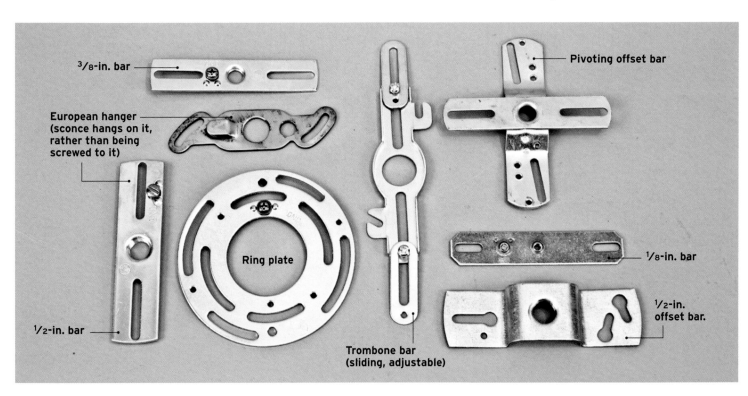

3/8-in. bar
Pivoting offset bar
European hanger (sconce hangs on it, rather than being screwed to it)
Ring plate
1/8-in. bar
1/2-in. bar
Trombone bar (sliding, adjustable)
1/2-in. offset bar.

What follows is an overview of how various fixtures mount to outlet boxes. Later in the chapter (see p. 85), we cover how to attach a standard octagonal outlet box for a ceiling fixture.

All metal brackets, boxes, and lamp fixtures must be grounded to be safe. The green grounding screws have a specified thread count to ensure a positive connection to metal boxes or plates.

Flat-mounting brackets

Typically, a mounting bracket screws to an outlet box, and the fixture attaches to the bracket, either by machine screws or, as is more common for chandeliers, by a threaded post that screws into a threaded hole in the center of the mounting bracket. Brackets can be as simple as a flat bar with screw slots; but some adjust by sliding, whereas others are offset slightly to provide a little more room for electrical connections—and fingers. Ring brackets can be rotated so the slots line up perfectly with outlet box and fixture screw holes.

Even simple brackets give you several mounting options. The flat bar shown in the top photo at right, for example, is slotted to receive fixture machine screws. It also receives a threaded nipple to which a chandelier will mount.

➡️ **For more on installing chandeliers, see p. 85.**

Matched brackets

Some mounting brackets are specifically matched to a fixture, as with the clever martini wall sconce shown on p. 84. Because the fixture designer didn't want screws on the face of the fixture base, he specified a flanged bracket, which receives screws on the side, where they'll be less visible.

No brackets

Some fixtures, such as the recessed lighting fixture shown on p. 88, don't require a mounting bracket. The fixture's can (housing) is its own junction box; and, once inserted into a hole cut in the ceiling, the fixture is supported by the ceiling it sits on. The fixture can is further secured by integral clips and trim pieces that pull it tight to the plaster or drywall ceiling. Recessed cans are IC rated (they may be covered with insulation) or non-IC-rated (cannot be covered with insulation). (*IC* stands for "insulated contact.")

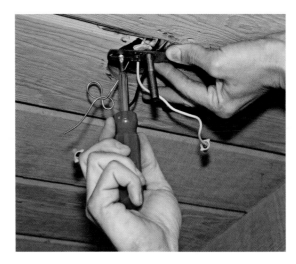

Screwing a mounting bracket **with a threaded nipple to a ceiling box.**

This wall sconce requires **a special flanged ring bracket to mount it.**

An IC-rated **incandescent fixture can be covered with insulation.**

⚠️ WARNING

Most bar and box assemblies are rated for 50 lb; if your light fixture weighs more than that, install a fan box instead.

INSTALLING A SIMPLE PORCELAIN FIXTURE

1 Test the hot wire for power.

2 Then test the neutral wire.

5 Connect the supply wires to the screw terminals.

6 Screw the fixture to the box.

Before replacing or installing any fixture, shut off the power to the outlet and test to be sure it's off. Use an inductance tester to test both hot **❶** and neutral **❷** wires for power because there's no guarantee that the fixture was wired correctly.

If the existing cable has fraying, fabric sheathing but the thermoplastic insulation around individual wires is intact, it's safe to reattach the wires to a new fixture. To keep the fabric from fraying further, wrap the sheathing with electrical tape. Use black tape on the hot (black) wire **❸** and white tape on the neutral (white) so there will be no future confusion about which wire is which **❹**. Of course, if the sheathing is intact, it's not necessary to wrap wires with tape.

Connect wires to the appropriate screw terminal on the fixture: neutral wire to the silver screw, hot wire to the brass screw **❺**. If a fixture has two or more mounting screws, start each screw before tightening any one screw all the way down. It will be easier to line up the screws to the holes in the outlet box or mounting bracket **❻**.

With porcelain fixtures, don't overtighten the screws that hold it to the wall or ceiling. Likewise, never overtighten a bulb. A snug fit is fine **❼**.

3 Wrap black tape on hot wire.

4 White tape marks the neutral wire.

7 Don't overtighten the bulb.

PLAY IT SAFE!

Before working on any electrical fixture, shut off power to the circuit by flipping the breaker or removing the fuse controlling that circuit. Then use an inductance tester to be sure the power is off.

Electrical codes require that all fixtures and devices—everything that gets installed—must be "listed" and must be recognized by the authority having jurisdiction (AHJ), usually the local inspector checking your installation. (Typically, light fixtures will have an Underwriters Laboratories [UL] listing.) If an inspector doesn't see a UL sticker, he or she could ask you to remove the fixture.

Pay attention to a fixture's wattage rating, usually specified on a sticker on the fixture's base. Substituting a bulb with a higher wattage can overheat and damage the fixture and, in some cases, ignite nearby combustible surfaces.

⚠ WARNING

Porcelain fixtures have endured for decades because they're inexpensive and reliable. But they are relatively fragile. Avoid overtightening the machine screws that hold a porcelain fixture base to a box, because if you fracture the base you'll have to replace it. When the screw head is just snug, stop turning.

REMOVING AN EXISTING WALL SCONCE

1 Support the shades as you unscrew them.

2 Test to make sure the power is off.

3 Unscrew the existing fixture.

4 Remove the fixture to expose the wires.

5 Disconnect the wires and remove the old bracket.

When removing an existing sconce, save all the old screws and incidental hardware—you never know what you might need when you attach the new one. Choose a sconce that suits your taste. In this case, we used something a bit more playful—a sconce whose shade is a martini glass and whose halogen bulb shines through colored glass ice cubes. Because the new sconce is a low-voltage unit, it also requires a transformer, which fits under the fixture base.

First remove the glass shade from the existing fixture; most are held on by small setscrews. Always support the shade—especially if it's inverted—to prevent its falling out and breaking **1**. Turn off power to the outlet, then use an inductance tester to make sure the power is off **2**. If the tester doesn't light, it's safe to disassemble the fixture and handle its wires.

Remove the fixture's mounting screws; in this case, they're on the side of the fixture base **3**. Pull the fixture away from the wall to reveal its mounting brackets and the wire connectors that splice the supply and fixture

wires **4**. (The green-and-yellow striped wire is the fixture's ground lead.)

After disconnecting the splices to the fixture leads, remove the special mounting bracket because it won't be needed to mount the new fixture **5**. (Its holes won't line up with the new fixture's mounting screws.) Save this bracket in case you want to reinstall the old sconce elsewhere.

TRADE SECRET
It's always best to turn off the circuit breaker feeding the fixture rather than relying on the light swithch to disconnect power.

CONNECTING A NEW SCONCE

Once the old sconce has been removed, you're ready to connect the new one. Install the new sconce's mounting bracket **❶**. In this case, the new bracket has tab brackets on each side rather than a continuous flange, like the old one.

If you hold the fixture one handed as shown, it frees your other hand to make connections **❷**. Here, the fixture takes a 12v, bi-pin halogen bulb, so it requires a transformer (the black box) to reduce the house's 120v power. (The transformer here is an electronic, solid-state device—essentially, a circuit board—with no moving parts.)

Splice the ground wires, then the neutral wires, and then the hot wires **❸**. After you finish splicing the wire groups, push the excess wire and the wire connectors into the outlet box behind the mounting bracket so there will be room for the transformer under the fixture base **❹**.

Line up the holes in the fixture base to the holes in the mounting bracket, and screw the base to the bracket **❺**. Don't tighten one screw all the way down until you've at least started the screw on the other side.

WHAT CAN GO WRONG

On occasion, a mounting bracket is smaller than the hole or the outlet box will have been set below the level of the drywall. To prevent the bracket's being drawn into the wall, put washers behind the screw holes of the bracket.

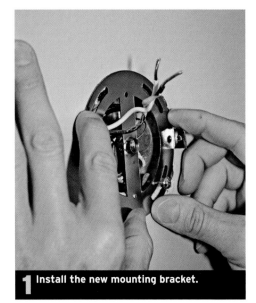

1 Install the new mounting bracket.

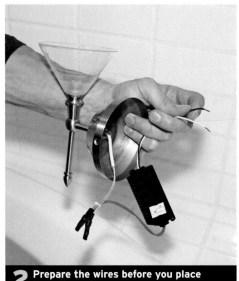

2 Prepare the wires before you place it on the wall.

3 Use wire connectors to splice wires.

4 The transformer fits under the base.

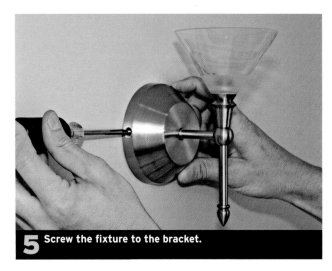

5 Screw the fixture to the bracket.

INSTALLING HALOGEN BULBS & ACCESSORIES

Before inserting any bulb into a fixture socket, check its rating. As noted earlier in this chapter, if the halogen bulb pins don't fit the socket, don't force them—you may have the wrong bulb. Grip the protective plastic wrapping—not the bulb—as you insert the halogen bulb into the socket ❶. As you remove the bulb from its packaging, gently pinch the end of the bulb—a little like squeezing a fast-food packet of ketchup—until its pins stick out through the plastic. Once you've pressed the bulb into the socket, you can easily slide the plastic off ❷.

Lo-vo bulb pins are so tiny that if they become oxidized, carbon can build up in the socket, causing the bulb to flicker or not shine at all. To avoid replacing a socket (which means rewiring the lamp), electricians routinely apply an antioxidant paste to the lo-vo pins before inserting them into a socket.

If your fixture has novelty items such as these glass ice cubes, take care when installing them so you don't damage the halogen bulb ❸. The cubes are tempered glass, so they can withstand heat. Always allow the lamp and cubes to cool before handling them or you'll burn yourself badly ❹.

1 Grip the plastic wrapping when inserting the bulb.

2 A bi-pin bulb must be properly seated in the socket.

3 Carefully place the glass cubes.

4 Party time!

Touching a halogen bulb with bare fingers shortens the bulb's life. Instead, slit the plastic bag the bulb is shipped in and grip the bag as you insert the bulb.

PREPPING & MOUNTING A CEILING BOX

1 Remove the center knockout.

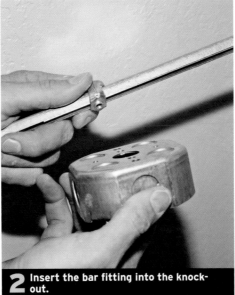

2 Insert the bar fitting into the knockout.

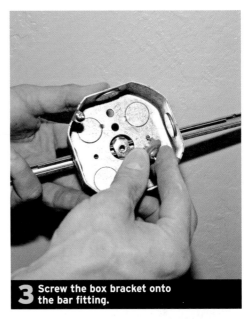

3 Screw the box bracket onto the bar fitting.

4 Bend bar tabs upward for retrofit applications.

5 Extend the bars and screw them to joists.

S tandard ceiling box and bar assemblies are rated for 50 lb. If your light fixture weighs more than that, install a fan box instead.

➤ **For more on installing fan boxes, see p. 147.**

Start by using a sturdy pair of needle-nose pliers or lineman's pliers to knock loose and twist out the center knockout in the box **1**. To attach the box to the bar, line up the threaded fitting on the bar to the knockout in the center of the box **2**.

Inside the box, there's a screw and a small bracket. Screw the bracket to the threaded fitting on the bar, which solidifies and strengthens the assembly **3**. Mounted together, the box and bar are rated to support a 50-lb. light fixture.

In new construction, this bar and box assembly would be installed before the ceilings are covered, from below. Thus the bar's tabs face down. In a retrofit, however, you'll be screwing or nailing the tabs to joists from above (if there's access). So when retrofit-

ting a ceiling box, bend up the tabs and they'll be easier to secure **4**.

Extend the bar until both ends are snug against the joists. Measure the thickness of the ceiling, adjust the height of the mounting bar so the box will be flush to the ceiling below, and screw the bar tabs to the joists **5**.

DISCONNECTING A CHANDELIER

1 Unscrew the canopy and remove it.

2 Test to make sure the power is off.

3 Remove the wire connectors.

4 Support the chandelier while removing the fixture.

5 If necessary, remove the existing bracket.

TRADE SECRET
If a chandelier canopy is on the small side, make sure the ceiling box is large enough to house the wires (and wire connectors) to be connected. Conversely, if the box is on the small side, a large canopy can hide the connections.

In the sequence shown here, the homeowners swapped a 1920s gothic chandelier for a large, handcrafted Mexican one. Both fixtures were quirky, and the canopy of the Mexican fixture was small for a fixture of its size. Fortunately, the ceiling box was deep enough to house the supply cable and the wires feeding the four chandelier arms. Had there been a shallow "pancake box" in the ceiling, the canopy would have been too small to house all the wire connections.

Because a helper didn't show up, the electrician had to support the very heavy fixture with one hand while he attached wire connectors with the other. Sometimes you've got to make do with the resources at hand. But if you install a chandelier this big or unwieldy, by all means get help. Before disconnecting a chandelier, turn off the breaker controlling the fixture. Unscrew the setscrew holding the canopy to the fixture post ❶ and slide the canopy down the post to expose the wire connections behind it.

After lowering the canopy, pull out the wire splices—being careful not to touch bare wires—and use an inductance tester to see if there's power ❷. If the tester glows, there's power: Flip the circuit breaker or remove the fuse controlling the circuit and test again. Once you're sure the power is off, twist off the wire connectors and separate the wires ❸. To avoid misplacing them, temporarily twist the wire connectors onto source wires.

Remove the chandelier. In most cases, that means unscrewing two mounting screws; but here the fixture post was threaded, so it was necessary to spin the whole fixture to unscrew the post from a threaded bracket in the box ❹.

See if the replacement chandelier can be mounted to the existing bracket; if not, remove the bracket ❺. Also, if the outlet box is not flush to the ceiling, now is the time to rectify that condition.

INSTALLING A CHANDELIER

Before installing a chandelier, preassemble the mounting bracket for the new fixture. This bracket is typical: A threaded nipple screws into the mounting bracket; a chandelier fitting will screw onto that nipple. Note, too, the green grounding screw, which will secure a pigtail from the ground-wire splice ❶. Then screw the mounting bracket to the ceiling box ❷.

Get help if the fixture's heavy, especially if it screws directly to the box rather than to a nipple. But if you must install it single-handedly, assemble all the tools and parts beforehand so you can focus on lifting the fixture and securing it quickly ❸. Make sure the wires won't obstruct the nipple in the box, and have the cap screw handy. As you raise the fixture, support its canopy ❹ rather than holding individual lamp supports, which may be assembled in sections and come apart.

If the nipple supporting the fixture is long enough, there will be adequate room to reach in and splice wires. Splice the ground wires first, then the neutrals, and then the hot wires ❺. When the splices are complete, tuck the conductors into the ceiling box or behind the canopy, hold the canopy flush to the ceiling, and tighten the mounting screws all the way down. Here, a cap nut covers the end of the threaded nipple ❻.

WARNING

When dealing with any fixture, disconnect the power at the service panel or fuse box before mounting the fixture. If the wires are jammed behind the canopy, there's a chance of nicking a wire and possibly creating a ground fault.

1 The mounting bracket is equipped with a grounding screw.

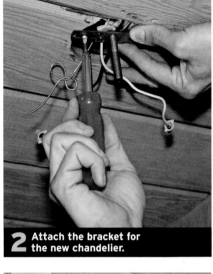

2 Attach the bracket for the new chandelier.

3 Look—and think—before you lift. What must you do first?

4 Support the canopy as you raise the fixture into place.

5 With the fixture supported, splice the wires.

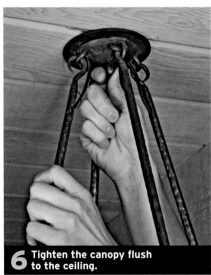

6 Tighten the canopy flush to the ceiling.

CUTTING A CEILING HOLE FOR RECESSED LIGHTING

RECESSED LIGHT FIXTURE

Recessed light fixtures vary. The low-voltage model in the photo sequences has a transformer at the end of its assembly to reduce house voltage. The drawing shows a model that runs on house voltage (120v), so it has no transformer. If the unit is watertight, it will have additional trim or lens elements. Closely follow the installation instructions provided with your fixture.

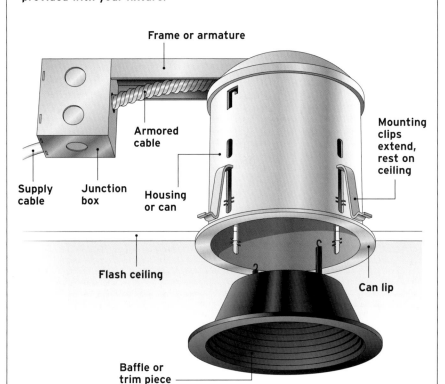

Frame or armature

Armored cable

Supply cable

Junction box

Housing or can

Mounting clips extend, rest on ceiling

Flash ceiling

Can lip

Baffle or trim piece

Retrofitting recessed lighting In retrofit installations, the supply cable to the recessed lighting unit typically comes from an existing ceiling box or nearby switch box. The supply cable feeds to an integral junction box on the fixture. Finding the nearest power source and fishing the wires to the fixture are always an adventure if there's not accessible space above. If the recessed fixture is a low-voltage unit, such as the one shown here, it will come with a transformer, which reduces the 120v current of the supply cable.

As the name implies, recessed lighting fixtures fit up into space above the ceiling and so don't disturb the flat plane below. Recessed fixtures distribute light evenly and thus are frequently used to illuminate work areas or tight spaces. (The watertight unit shown here is installed in a shower alcove.) Because a retrofit recessed light can (housing) doesn't weigh much, it rests lightly atop ceiling drywall or plaster. Spring-loaded clips or trim pieces draw the can lip tight to the ceiling surface. Of course, all connections are done with the power off.

There's no absolute on where to place a recessed light; but in a small space, such as a shower alcove, a fixture centered in one direction or another will look best ❶. In addition, you may want to use a stud finder to avoid hitting ceiling joists above.

Drill a pilot hole to see what's above ❷, and to make sure there's room for the can. Make the hole small because if there's an obstruction above it, you'll need to patch it. After drilling the hole, you can insert a 4-in. piece of bent wire and rotate it to see if it hits a ceiling joist. Also, drill a small pilot hole to keep the point of a hole-saw blade from drifting.

Keep the drill vertical, and the circle of the sawblade parallel to the ceiling ❸. There are special carbide hole saws for drilling through plaster. A bimetal hole saw will also cut through drywall or plaster, but it'll destroy the saw in the process. Wear goggles.

If the hole saw is the right size for the can, you won't need to enlarge it. But for the light shown here, the saw was a shade too small, so the installer used a jab saw to enlarge the hole slightly ❹. In a pinch, you can also use just a jab saw.

Test-fit the unit ❺. Although you want the can to fit snugly, the unit's junction box and transformer also need to fit through. The black box about to enter the hole is the transformer.

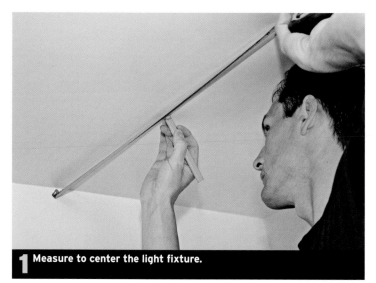

1 Measure to center the light fixture.

2 Drill a pilot hole.

3 Keep the drill aligned vertically.

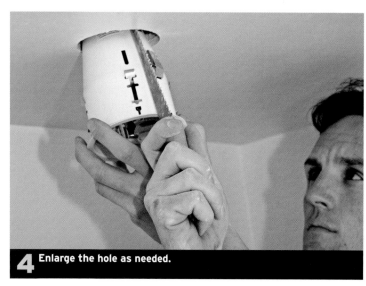

4 Enlarge the hole as needed.

5 Test-fit the fixture.

TRADE SECRET

To minimize hitting pipes or cables when drilling into a ceiling, use a bit that's 1 in. long—just long enough to drill through the plaster or drywall. Use a cordless drill to further reduce the chances of shocks.

WIRING A RECESSED FIXTURE

To wire a recessed fixture, remove a Romex knockout from the unit's integral junction box ❶. Inside the knockout, there is a spring-loaded, strain-relief clamp that will grip the incoming cable, so you don't need to insert a Romex connector. Run a length of (unconnected) Romex cable from the nearest power source and feed it into the knockout just removed. (To wire the box with AC or MC cable, remove one the circular 1/2-in. knockouts and insert an appropriate connector.)

➡ **For more on connectors, see p. 36.**

Inside the fixture's junction box will be two sets of wires that were spliced at the factory. They connect the secondary wires that run from the transformer to the socket. (At the transformer, the current is reduced from 120v to 12v, so polarity is no longer an issue.) There are also three unconnected fixture leads in the box, to which you'll splice the supply wires ❷. Using wire connectors, connect the incoming ground wire to the green fixture lead, the incoming neutral to the white lead, and the hot wire to the black fixture lead ❸.

Tuck the spliced wire groups into the fixture junction box ❹. At the right of the photo is a piece of threaded rod that can be adjusted to support the transformer at the correct height. Snap shut the junction box cover ❺. As with other outlet boxes, Code determines the number of wires you can splice in a fixture junction box, based on the cubic inches in the box.

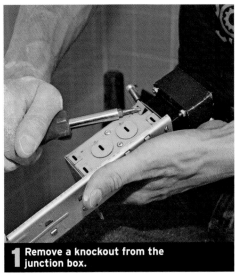

1 Remove a knockout from the junction box.

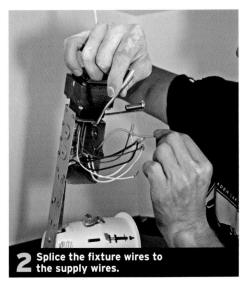

2 Splice the fixture wires to the supply wires.

3 Connect the ground wires first.

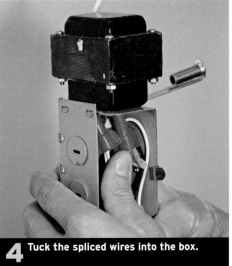

4 Tuck the spliced wires into the box.

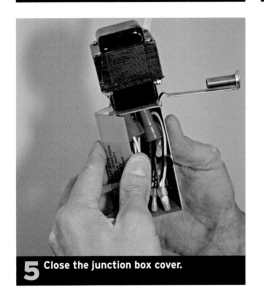

5 Close the junction box cover.

TRADE SECRET

If you have access to the space above, staple the Romex to a joist, leaving at least 12 in. of free cable so you can push the fixture components into the hole. A cable that's too short may prevent this.

SECURING THE CAN

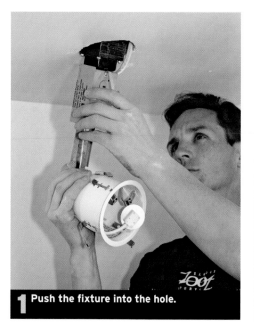

1 Push the fixture into the hole.

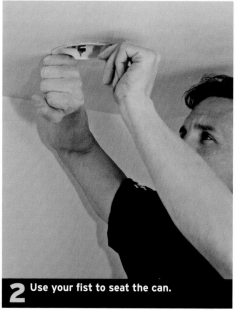

2 Use your fist to seat the can.

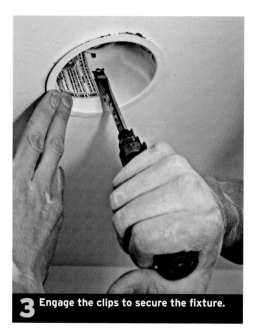

3 Engage the clips to secure the fixture.

4 Insert the bulb with steady pressure

5 Install the trim assembly.

TRADE SECRET

The pros apply a tiny dab of antioxidant paste to the lamp pins before seating them in a fixture socket to prevent oxidation. A good practice, but not imperative.

Once the recessed lighting fixture has been wired, push the fixture into the hole, being careful not to bind the Romex cable as you do so **1**. If the fit is snug, use the side of your fist to seat the lip of the fixture flush to the ceiling **2**.

Use a screwdriver to push up the spring-loaded clips that pivot and press against the back face of the drywall to hold the fixture snugly in place **3**. To remove the fixture later, pop out the clips.

Insert the bulb (the fixture here uses, an MR-16 bi-pin halogen bulb) into the fixture socket **4**. Here, the installer is gripping the lamp's reflector, not the bulb itself. The lamp pins should seat securely. Install the trim piece—this one has a watertight gasket. Snap in the lamp and socket and push the assembly up into the can. The three arms on the side of the assembly will grip the inside of the can **5**.

UNDERCABINET FLUORESCENT FIXTURES

Position undercabinet fluorescent fixtures to give you the countertop light coverage you want in each area. If you're running Romex cable to the fixture, you must protect it, either by placing the fixture flush against the back of the cabinet or by covering the Romex with a piece of trim. If you prefer to place the fixture out from the wall, you can also run MC cable, which can be left exposed.

➡ **For more on MC cable, see p. 30.**

Run the supply cable to the fixture location. Allow at least 12 in. of cable inside the fixture to facilitate wire splices. Remove the sheathing and then feed the supply wires through a knockout in back of the housing shell. Clamp the cable, then screw the shell to the underside of the cabinet ❶. Hold up the shell with one hand and use a cordless screw gun in the other.

Snap on the wiring compartment ❷, which contains numerous prewired connections, and the ballast (the white rectangle seen at the right in the photos). Use wire connectors to splice the fixture leads to the supply wires ❸. (Extend the stranded wire ends beyond the ends of the solid wires so that the wire connectors engage the stranded wires first.)

➡ **For more on splicing stranded wire, see p. 37.**

As you shut the wiring compartment, do your best to keep the wires neat and compact ❹. Messy wires are more likely to get pinched by the cover and shorting out—which means taking the unit apart and starting all over. Also, avoid locating the wire connectors too near the ballast, which may keep the cover from shutting. Screw in the cover screws that secure the wiring compartment.

As with most fluorescent tubes, insert their end pins into the slots in each keystone, then rotate the bulb clockwise to seat the pins in the sockets ❺. When turning fluorescent bulbs, grab them close to their ends and never force the lamps into place—they should seat easily. Finally, snap the plastic diffuser (cover) of the fixture into place. Be gentle because the diffusers break easily ❻.

UNDERCABINET FLUORESCENT FIXTURE

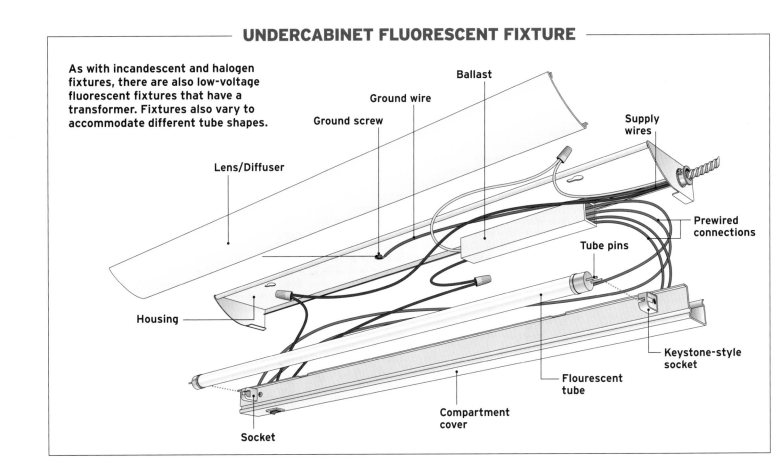

As with incandescent and halogen fixtures, there are also low-voltage fluorescent fixtures that have a transformer. Fixtures also vary to accommodate different tube shapes.

Ballast

Ground wire

Ground screw

Supply wires

Lens/Diffuser

Prewired connections

Tube pins

Housing

Keystone-style socket

Flourescent tube

Socket

Compartment cover

1 Feed in the wires, then screw the housing to the cabinet.

2 Snap on the wiring compartment.

3 Splice the fixture and supply wires.

4 Close the wiring compartment.

5 Slide the pins into the sockets, then rotate.

6 Snap the lens or diffuser into place.

LOW-VOLTAGE SYSTEMS

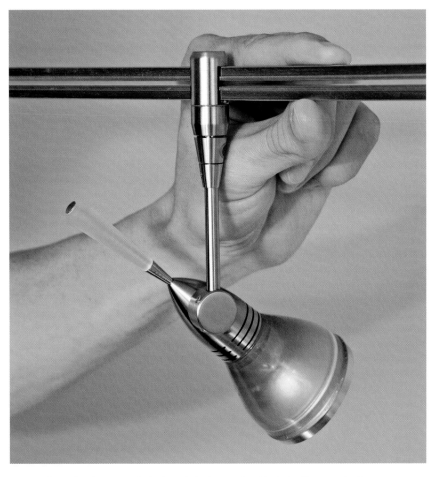

Low-voltage tracks **are safe to touch, but you should shut off power to the circuit anyway.**

Safety and low-voltage systems

Experienced electricians often handle low-voltage tracks while those parts are energized—with 12v of current—downstream from the transformer. If you install your lo-vo system correctly, and are careful *never* to touch the supply wires that run to the transformer, you can touch energized lo-vo tracks and install light fixtures without getting shocked. But if you have *any* confusion about which wires are 120v and which are 12v, shut the power off before doing any work on your lo-vo system. Best to be too cautious . . . and stay safe.

Low-voltage lighting systems are installed inside and outside houses and typically operate on 12v current, so they require a transformer to reduce standard house voltage from 120v. Transformers vary—some are coil-wound magnets, whereas others are electronic—so follow the installation instructions provided with your unit.

For more on low-voltage systems, see p. 235.

Lo-vo systems have become extremely popular because they're safer (12v is roughly the same voltage as a car battery) and

energy efficient; plus, their bulbs are easier to direct, enabling you to highlight a painting or a work area. Lo-vo bulbs come in a greater range of options than do incandescent bulbs; and, in general, lo-vo bulbs offer a better quality of light. Lo-vo systems can be controlled by standard switches or dimmers, but you should check the literature that comes with your system.

Though it's safe to touch the tracks of a lo-vo system, you should turn off the power when working on the system. The upstream part of the system (between the transformer and the power source) has

120v power, which could deliver a fatal shock. On the low-voltage side, there is a potential to short the system and damage the transformer.

After installing all the parts and doing a preliminary check of the system, it's safe to energize the system. *Note:* Because any high-intensity bulb can get very hot, do not install any cable fixtures within 4 in. of a combustible surface.) Track-lighting systems are inherently complex, so read the instructions carefully before you begin the installation.

INSTALLING LOW-VOLTAGE CABLE LIGHTING

Track lighting systems may have a remote transformer or a surface-mounted transformer. There is also a range of cable standoff supports, both rigid and adjustable, mounting to walls and ceilings; they must be mounted solidly to framing if they are to keep position when the cable is tensioned. Cables are typically spaced 4½ in. or 8 in. apart. The illustration has been adapted from installation instructions for a product from Alfa Lighting Systems; your instructions may vary.

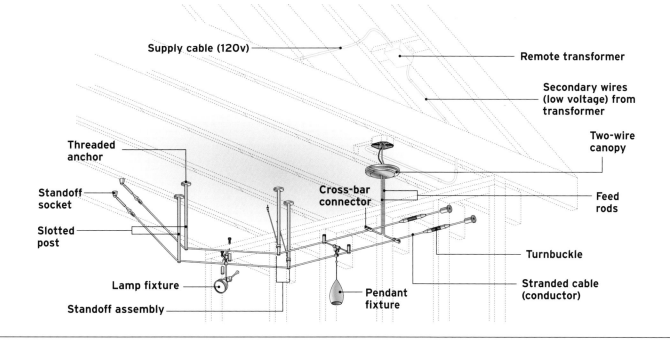

ANCHORING LOW-VOLTAGE STANDOFFS

The key to a good-looking, correctly functioning cable system is getting the cables taut and level. Thus your first task in installing a lo-vo system is finding solid locations in which to anchor the cables. In this installation, the principle anchors were *standoffs* in each corner so the cable could be stretched around the perimeter of the room. Alternatively, you can anchor standoffs in a ceiling to support cables and bulbs. Standoffs are also called *rerouters* because cables often change direction as they emerge.

Use a laser level to establish level anchoring points around the room. Predrill holes in the plaster for anchor screws ❶. Plaster is harder than drywall and there may be lath nails in the way, so wear goggles and have extra drill bits on hand. >> >> >>

1 Predrill all anchor holes.

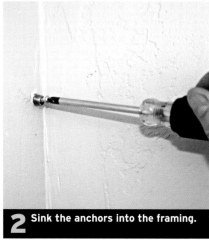

2 Sink the anchors into the framing.

ANCHORING LOW-VOLTAGE STANDOFFS (CONTINUED)

Because cables will be stretched taut, standoff anchors must be screwed to wood framing—in this case, into doubled studs in the corner ❷. Use screws at least 2 in. long to attach the anchors. After sliding a chrome base plate (washer) over the anchor, screw the standoff socket to the exterior threading of the anchor ❸. Insert the ball end of the fiberglass rod into the standoff socket ❹. This ball-and-socket assembly allows the standoff to swivel freely so you can fine-tune the cable positions. The cables will be spaced 4½ in. apart.

3 Screw on the standoff socket.

4 Insert the standoff into the socket.

RUNNING CABLE

The next step in installing this system is to run the cable. Note the standoff at the top of photo ❶. Because the walls in this room were only 12 ft. apart, both cables could be run through a single angled standoff with two slotted posts. If the walls are farther apart, you may need a cable support in the middle of the run.

Measure the cable length you need and cut it 3 in. to 4 in. long so you can insert the ends into turnbuckles without having to struggle. Once you tighten the turnbuckles and tension the cable you can snip off any excess cable. On the other hand, if you cut a cable too short, you'll have to discard it and start again with a new piece.

Place each cable into a slotted post, rather than pulling it through the slot, whose sharp edges can cut into stranded cable ❷. (This is a quirk of the particular system shown here; other standoff types allow you to pull cable more freely.)

After placing the cable into a slotted post, screw on the post cap to keep it from popping out when the cable is tensioned ❸. As you place the cable into the subsequent standoffs, loosely tension it to take up the slack ❹. With this system, the second cable will be about 4½ in. from the first.

1 Cut the cable a bit long.

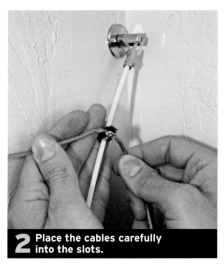

2 Place the cables carefully into the slots.

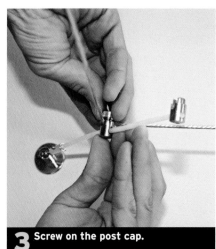

3 Screw on the post cap.

4 Tension the cable to take up the slack.

TENSIONING CABLES

Correctly installed, the cables of a low-voltage system should be more or less horizontal and equally spaced (parallel along their length). As with most systems, this installation uses turnbuckles to tighten the cables after they have been placed in the standoffs. The cables' Kevlar® core prevents stretching or sagging once the lightweight fixtures have been installed.

Once you've loosely run the cable, insert an end into a turnbuckle ❶. The cable end feeds in the end of the turnbuckle and exits in a slot in the middle. Tighten the setscrew(s) on the assembled turnbuckle to keep the cable from pulling out, then trim the excess cable sticking out ❷. Don't rush trimming the cable: Wait until you've made final adjustments to the whole layout before trimming.

A turnbuckle's center post has a thumbscrew with threads on both ends. As you turn the thumbscrew in one direction, it draws tight both ends of a cable; turn in the opposite direction to slacken cable tension ❸. With cable attached to both sides of a turnbuckle, the installer struggles to draw the cable tight enough to join them ❹ This is a good reason not to trim the cable until the turnbuckle starts tightening both ends.

A certain amount of adjustment is necessary after both cables are taut. Here, the installer adjusts the cables so that the standoff comes out of the corner at a 45-degree angle, thus ensuring that the wires will be equidistant ❺.

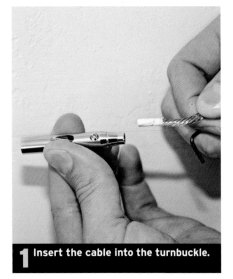
1 Insert the cable into the turnbuckle.

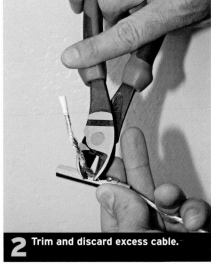
2 Trim and discard excess cable.

3 One section of the turnbuckle threads into the

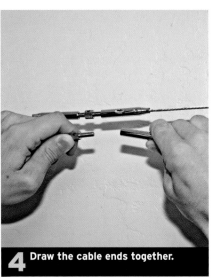
4 Draw the cable ends together.

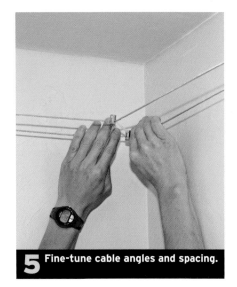
5 Fine-tune cable angles and spacing.

TRADE SECRET

When working with stranded cable, tape the cable ends and then cut in the middle of the tape to keep the strands from unraveling.

LOCATING THE CANOPY

Once the cable has been installed, you'll need to locate the canopy. The canopy (also called a *canopy feed*) receives low-voltage current from the transformer and delivers it to the cables. The canopy mounts to a junction box on the ceiling or wall and should fit flush to the finish surface.

Before installing the canopy, make its holes—or those of its mounting bracket—line up to the holes of the junction box you'll be installing in the wall or ceiling ❶. The slotted mounting bar on the back of this two-wire canopy can fit several box widths.

If the canopy will be ceiling mounted, retrofit a hole for it. To minimize the mess, use a hole-cutting tool with a dust cover ❷. *Note:* A screw gun with a ½-in. chuck will accept large-shank tools such as the one shown here. Set the hole-cutting tool's blade to the diameter of the junction box ❸. For large holes, this tool has a counterweight that attaches to the right side of the cutting bar to balance the torque of the blade.

Hold the cover of the tool snug against the ceiling so it can contain the dust ❹. Wait a few seconds for the dust to settle inside the cover before lowering the tool.

TRADE SECRET

Fixture canopies are often polished chrome, which is easy to cloud with fingerprints. Wearing lightweight plastic gloves solves both problems; nitrile plastic gloves are especially flexible.

1 Make sure screw holes line up.

2 Use a dust cover to minimize the mess.

3 Set the cutter to the diameter of the box.

4 Hold the cover snug against ceiling as you drill.

MOUNTING THE BOX & WIRING THE TRANSFORMER

1 Mount the ceiling box.

2 Place the transformer above the attic insulation.

3 Insert the lo-vo cable into the ceiling box.

4 Insert the cable into the transformer.

5 Run the supply cable to the transformer.

After the canopy has been located, it's time to mount the box and wire the transformer. The transformer shown here permits 12v or 24v wiring. Your installation may vary, so follow the instructions provided.

When retrofitting a ceiling box, bend up the bar tabs to make them easier to nail. Extend the support bar until its tabs are snug against joists, adjust the height of the bar so the box is flush to the ceiling, and then screw the tabs to joists **1**.

➤ For more on installing ceiling boxes, see p. 85.

Install the transformer above the insulation so that its vents work properly and the unit can be accessed easily **2**. Run the secondary (lo-vo) cable between the transformer and the ceiling box, stapling it within 12 in. of a box and every 4 ft. it along its run **3**. Feed the other end of the lo-vo cable into the transformer **4**. The gray cable connectors are one-way clamps: easy to insert but difficult to pull out.

Run the supply cable (120v) from an existing outlet to the transformer **5**. Protect the cable by stapling it to the side of a joist or to a runner added for the purpose. >> >> >>

MOUNTING THE BOX & WIRING THE TRANSFORMER (CONTINUED)

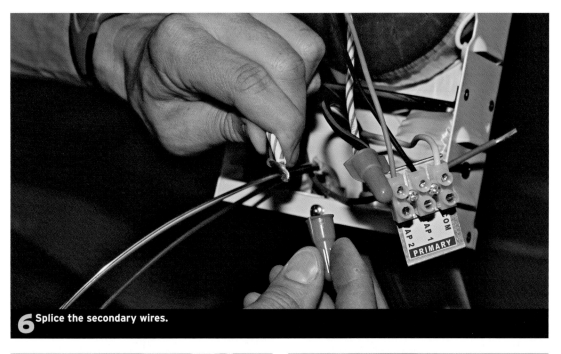

6 Splice the secondary wires.

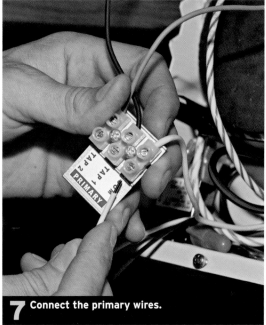

7 Connect the primary wires.

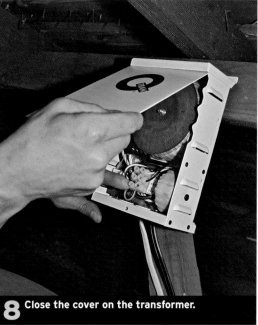

8 Close the cover on the transformer.

After stripping ½ in. of insulation off the wire ends, use wire connectors to splice the secondary wires, which run from the transformer to the ceiling box **6**. Next, splice the primary ground wires, using a wire connector. Then connect the primary neutral to the common (neutral) terminal, as shown, and the primary hot wire to one of the tap terminals. Tighten the terminal screws to grip the wires **7**.

Close the transformer cover to protect the connections inside **8**.

INSTALLING THE CANOPY

The canopy is installed after the transformer has been wired. Install the mounting bracket to the new ceiling box; it will support the canopy that supplies lo-vo power to the cables ❶. The wire hanging from the box is the secondary (lo-vo) cable from the transformer.

Separate and strip the two wires in the lo-vo cable and solder their ends. Soldering fine-strand wire makes it solid and unlikely to smash flat as you tighten down the set screws on the canopy terminals. Soldered wire is also less likely to arc and overhead ❷. (Note the tiny Allen wrench inserted into the setscrew on the right of the photo.)

Use mounting screws that are long enough to extend beyond the canopy face ❸; they're faster to install than short screws because they give you room to maneuver.

Slide the canopy over the mounting screws and turn the canopy cap nuts onto the mounting screws ❹. When the nuts bottom out on the screws, continue turning the cap nuts, which will turn the extra-long screws back up into the box. This will make the canopy nice and snug.

1 Attach the mounting bracket to the ceiling box.

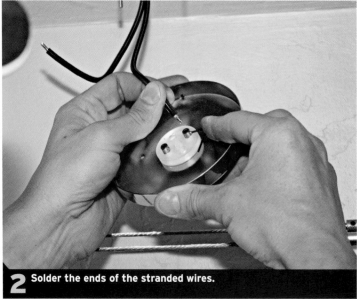
2 Solder the ends of the stranded wires.

3 Using extra-long mounting screws eases the installation.

4 Install the cap nuts on the mounting screws.

ATTACHING THE FEED RODS & FIXTURES

1 Bend the feed rods into position.

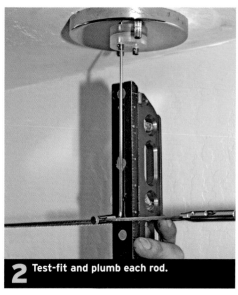

2 Test-fit and plumb each rod.

3 One rod has already been attached; the other has not.

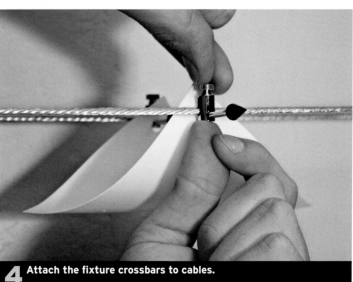

4 Attach the fixture crossbars to cables.

5 Once everything is installed, turn on the power.

The final task for installing a low-voltage system is adding the feed rods and fixtures. The feed rods transfer low-voltage current from the canopy terminals to the cables. Setscrews on the terminals secure the rods.

Because the canopy is centered over the two cables, bend both feed rods so they'll be an equal distance from the cables **1**. This task takes strong hands and pliers that won't scar the chrome finish of the rods. If you secure one end of the rod in a vise, put wood scraps between vise jaws and the rod to prevent scarring.

Bend the first rod and test-fit it, using a torpedo level to ensure that the rod is plumb and the cable is still level **2**. If the first rod fits well, use it as a template for the second. Repeat the process with the second feed rod. Note that rod ends are slotted like the standoff posts that anchor the cables in the corners of the room. A cap nut screws on to the slotted rod end to capture the cable **3**.

Insert the bulbs into the fixtures before you attach fixtures to the cables. If your system uses halogen bulbs, don't touch them with your bare hands because the oil in your skin will shorten the bulb life. In general, fixtures need to connect both cables to become energized, so each fixture has a crossbar that spans the cables. Hand-tighten the fixture connectors so they're snug **4**. After you'll installed all the fixtures and surveyed the system, turn on the power **5**.

INSTALLING LOW-VOLTAGE MONORAIL TRACK LIGHTING

In the installation shown, the transformer is surface-mounted in a circular housing. Although the transformer housing must be mounted to a ceiling box to be adequately supported, individual standoff supports can be anchored in drywall or plaster alone because the track and fixtures are lightweight. The illustrations have been adapted from installation instructions from a product by Tech Lighting℠; your instructions may be different.

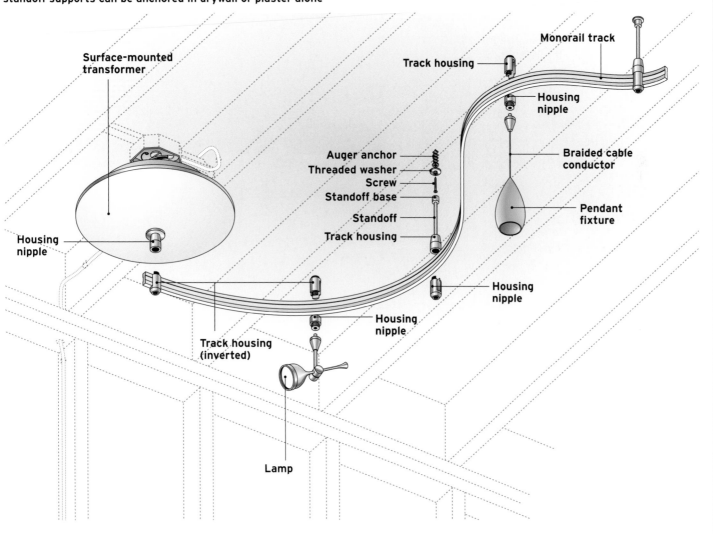

- Surface-mounted transformer
- Monorail track
- Track housing
- Housing nipple
- Auger anchor
- Threaded washer
- Screw
- Standoff base
- Standoff
- Track housing
- Braided cable conductor
- Pendant fixture
- Housing nipple
- Housing nipple
- Housing nipple
- Track housing (inverted)
- Housing nipple
- Lamp

Before installing lo-vo monorail track lighting, turn off the power at the breaker panel or fuse box. As with any low-voltage system, a transformer will reduce the 120v current to 12v or 24v. Because monorail track systems have a lot of small parts that are easily lost, don't open the packages until you're ready to use the parts.

In cross-section, the monorail track is two square pieces of chrome-plated copper conductor sandwiched together with a clear (insulating) plastic piece. Although the track is sturdy, it can be bent freehand or shaped against a curved piece of plywood.

LOCATING THE TRACK & STANDOFFS

If possible, before working overhead, shape the track sections of your monorail system and preassemble them on the floor, then use a plumb bob or laser level to plumb up to the ceiling. If there's an obstruction in the room you'll have to improvise. First determine where you'll place the transformer box for the system, then mark off the standoffs that will mount the track to the ceiling.

With a helper holding one end, hold sections of the track against the ceiling and mark off standoff points at the track ends, where sections meet, and at intervals suggested by the fixture maker—usually, every 3 ft. ❶. Standoffs have multiple parts, which anchor them to the ceiling and support the track (See "Standoff Parts" below).

Fortunately, because the track weighs little, you can mount standoffs almost anywhere on a drywall or plaster ceiling, using *auger anchors* with wide threads. (That is, you don't need to mount standoffs to framing.) Sink the anchors flush, then screw a threaded washer to each anchor ❷. Then, using a setscrew, screw the standoff base to that threaded washer ❸.

Mount all the standoffs so that when you raise the monorail again, you can attach it quickly to the standoffs and to the bottom of the transformer housing. If you need to reposition a standoff or two, it's easy to patch the small holes left by misplaced standoff anchors.

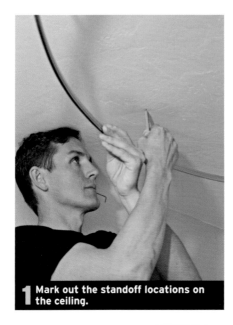

1 Mark out the standoff locations on the ceiling.

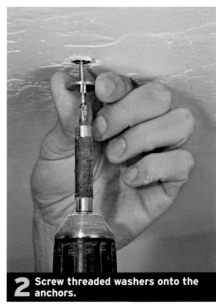

2 Screw threaded washers onto the anchors.

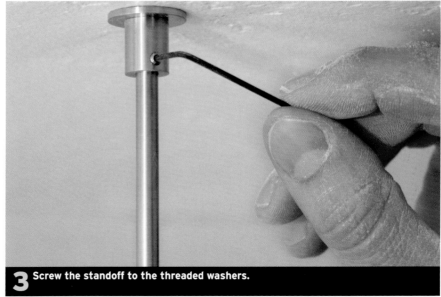

3 Screw the standoff to the threaded washers.

STANDOFF PARTS

Standoff supports for monorail track systems consist of many small parts.

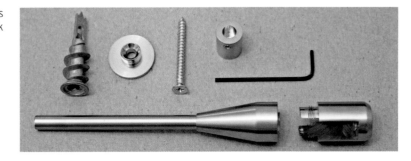

Clockwise, from top left: auger anchor, threaded washer, screw, standoff post, Allen wrench, housing, housing nipple.

MOUNTING & WIRING THE TRANSFORMER

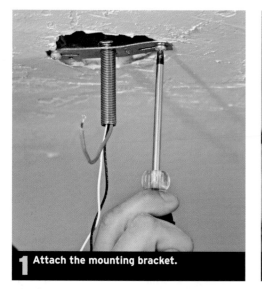

1 Attach the mounting bracket.

2 Feed the supply wires into the housing.

3 Secure the transformer housing.

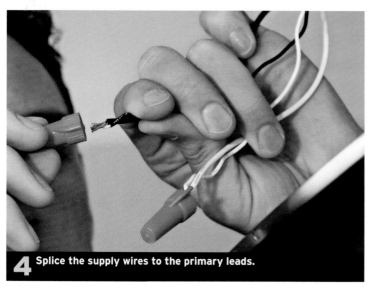

4 Splice the supply wires to the primary leads.

5 Test the connections and then close the housing.

Once the standoffs are installed, mount the ceiling box to a ceiling joist and attach the bracket that will support the transformer **1**. In this case, a nipple screws into the bracket and runs through the center of the transformer housing. The supply wires sticking out of the box will connect to the primary leads on the transformer.

➡ **For more on mounting boxes, see p. 85.**

Feed the supply wires into the center of the transformer housing **2**, push the top piece of the housing snug against the ceiling box, slip a washer over the end of the nipple, and then tighten the inside nut that secures the transformer housing **3**. The circular mass inside the housing is a magnetic transformer, which has a series of copper coils.

Many transformers come with secondary leads preattached, so that the installer need only splice supply wires to the primary fixture leads. Using the wire connectors provided, splice the ground wires first, then the neutral leads, and finally hot wires **4**. When all the wire connections are snug—gently tug on spliced wires to be sure—close the transformer housing by snapping the bottom to the top and tightening the setscrews provided **5**. (By the way, the fat striped wires are secondary leads, which run from the transformer to the lo-vo power feed that energizes the track.)

HANGING THE TRACK

With the transformer secured and the standoffs installed in the ceiling, you're ready to hang the track. Get help supporting the track until you have two or three points secured. Place the track into the standoff housing—the lowest piece on that assembly—then screw on the housing nipple ❶. Once that's done, tighten the setscrew that holds the nipple to the support rail.

Where track sections meet, join them with conductive connectors and support the junction with a standoff ❷. Once the track is supported at several points, loosely attach the housing and nipple assemblies at several points, slide them beneath the support rails, and use an Allen wrench to tighten the setscrews that join the nipples to the rails ❸. Make sure the connections are tight

Because the housing on the bottom of the transformer has setscrews on the side and the bottom, you'll need two different sizes of Allen wrenches. Loosen the small screw on the side of the housing nipple so you can turn it onto the threads of the housing. Once the nipple is tight on the housing, retighten that small screw. Then use a 3/16-in. Allen wrench to tighten the setscrew on the bottom of the housing ❹. Finally, retighten all connections and place end caps on the ends of the track ❺.

1 Place the track in the housing.

2 Use conductive connectors to join the sections.

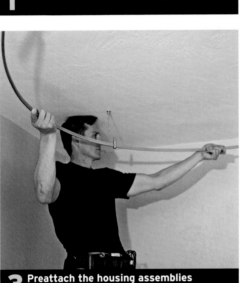

3 Preattach the housing assemblies on the track.

4 The transformer support requires two wrenches.

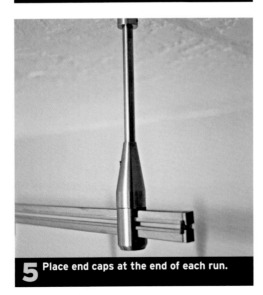

5 Place end caps at the end of each run.

TRADE SECRET
If you can't find anyone to help you support the track, bend short lengths of coat hanger into Z-shaped hanger brackets. Drill 1/4-in. holes into the ceiling and insert one end of the bracket; the other end supports the track. Once the track is up, remove the hangers and patch the holes with joint compound.

INSTALLING THE LIGHT FIXTURES

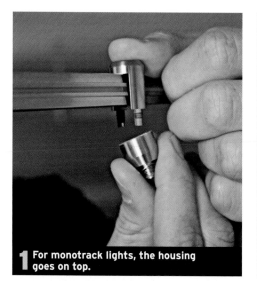

1 For monotrack lights, the housing goes on top.

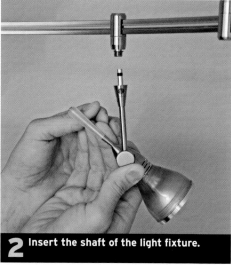

2 Insert the shaft of the light fixture.

3 Support pendant lights with braided cable.

4 Tighten the light supports on the pendants.

5 Energize the system for 20 minutes, then make sure the connections aren't overheated.

Once the monorail tracks are installed, you can add the light fixtures. The fixtures require a two-part assembly that straddles the monorail track; note that, in this case, the housing and nipples are inverted. Insert the housing on top of the track and the nipple on the bottom **1**. Screw the pieces together, insert the shaft of the light fixture into the inverted nipple **2**, then screw the fixture nipple onto the threaded housing nipple and hand tighten it.

Attaching the pendant lights is similar, although the light pendants use braided cable rather than a solid shaft. Braided cable can be shortened if necessary, allowing you to install the pendant lights at the same or varying heights **3**. Once you've adjusted the pendant cables, tighten the fixture nipple to the housing nipple on the track **4**.

Most manufacturers recommend energizing the system and then turning the lights on for 20 minutes or 30 minutes before checking the monorail connections to see if any are hot to the touch. Warm is normal, but hot connections should be retightened **5**. Do not, however, touch any halogen lights; they are certain to be hot enough to burn you (which is normal).

TRADE SECRET
Retighten all connections after the first 10-20 hours of use.

REPAIRING LOW-VOLTAGE CONTROLS

REPAIRING LOW-VOLTAGE controls will be familiar territory if you've read the earlier chapters on switches and installing light fixtures. Basically, doorbell buttons and thermostats are switches. To repair or replace doorbell units, thermostats, and transformers, you'll need only basic tools such as those needed to install switches or lighting. And like low-voltage lighting, doorbell units and thermostats are energized by a transformer, which reduces house voltage. You can safely test or handle existing low-voltage wires without turning off the power. But because transformers are energized with 120v house current, you must always turn off the power before testing or repairing a transformer or wire upstream—those that run from the transformer to a power source. Thus an inductance tester and a multimeter are essential tools to keep you safe.

DOORBELLS

Troubleshooting a doorbell, p. 110
Replacing a doorbell switch, p. 113
Installing a chime unit, p. 114
Installing a doorbell transformer, p. 115

THERMOSTATS

Testing an old thermostat, p. 116

Installing a programmable thermostat, p. 117

FURNACES

Replacing a furnace transformer, p. 118

TROUBLESHOOTING A DOORBELL

Troubleshooting a doorbell system takes a little detective work. On older systems, the problem is usually the chime or bell unit—we'll use the term *chimes* to denote either. In many cases, the plunger springs on the chimes become compressed, resulting in chimes that no longer ring predictably–or at all. Corrosion or dust buildup can also silence chimes. Try the tests given here to determine whether the problem is the switch, the chimes, the transformer, or the wiring between the transformer and the switch.

Start by testing the doorbell switch. Unscrew it and gently pull it out from the wall to expose its connections on the back side ❶. Disconnect one of the wires and, using a continuity tester, touch the tester clip to one switch terminal and the tester point to the other. Press the button: If the tester lights as you depress the button, the switch works ❷. If not, the switch is faulty and should be replaced.

A more thorough testing

Alternatively, you can detach both wires from the switch and touch their bare ends together to perform several tests. If the chimes sound when you join the wires, the switch is defective ❸. If the switch wires spark when you touch them but the chimes don't ring, test the chimes as shown in "Testing Chimes," on p. 112.

If there's no spark when you touch the switch wires, test the transformer instead ❹.

If the transformer is working, you'll need to replace the wiring to the doorbell switch or install a wireless system.

TEST THE SWITCH

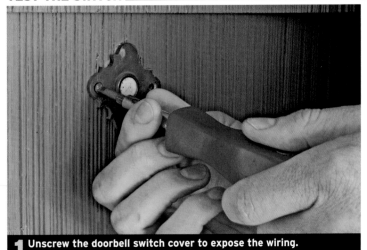

1 Unscrew the doorbell switch cover to expose the wiring.

2 Use a continuity tester to test the switch.

TEST MULTIPLE COMPONENTS

3 Join the two bare ends of wire to test the system.

4 Testing the transformer is the last option.

Perhaps the trickiest part of low-voltage repairs is running the cable. If you're replacing dead lo-vo wires, such as those running to a doorbell, try to twist and tape new low-voltage wires to the old ones and pull them through the walls. Because low-voltage wire is thin, it's flexible and much easier to pull than Romex.

If you must drill through an exterior wall to bring low-voltage wiring into a room, avoid hitting cables that may be hidden in the walls. The safest place to drill is generally low on the wall, away from studs.

→ **See "Rough-in sequence" p. 199.**

As shown in the rough-in sequences on p. 199, most house circuits are routed horizontally about 2 ft. high and secured to studs before entering outlet boxes. If you drill below that height and between studs, chances are good that you'll avoid hitting cables. To further reduce any chance of getting a shock, use a cordless drill for this task.

Once you've found the right location, drill downward to keep water from running into the walls.

→ **See photo 2, p. 125.**

After pulling the cable, fill the hole with siliconized latex caulk, which, unlike pure silicon, is paintable. When running low-voltage wire, staple it at least every foot to prevent sagging, strains on connections, and an unsightly appearance.

SINGLE DOORBELL WIRING

In a single doorbell system, the circuit runs from the transformer to the doorbell switch, from the switch to the chime unit, and then back to the transformer. By pressing the doorbell, you complete the circuit, and the chime unit rings.

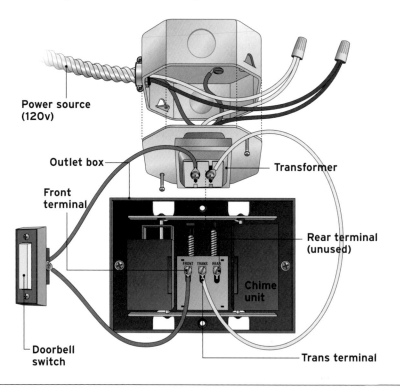

Power source (120v)

Outlet box

Front terminal

Doorbell switch

Transformer

Rear terminal (unused)

Chime unit

Trans terminal

FRONT TRANS REAR

WARNING

It's safe to handle the energized low-voltage wires that run from the transformer to the doorbell switch or from the chimes to the switch. However, because 120v current can harm you, turn off the power before testing or repairing a transformer or the wires upstream—that is, toward the power source—as shown in the upper right drawing.

TROUBLESHOOTING A DOORBELL (CONTINUED)

DOUBLE DOORBELL WIRING

A double doorbell system has two circuits, each controlled by a doorbell switch, whose power comes from the transformer. Thus the transformer for a double doorbell has three terminals. Typically, the chime unit has two different ring patterns (*ding* and *ding-dong*) so you can tell whether the visitor is at the front or rear door.

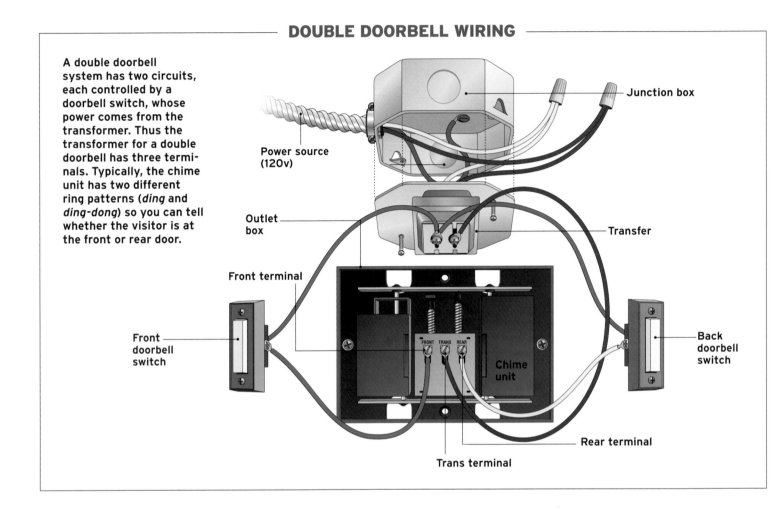

Junction box

Power source (120v)

Outlet box

Transfer

Front terminal

Front doorbell switch

Back doorbell switch

FRONT TRANS REAR

Chime unit

Rear terminal

Trans terminal

Testing chimes If the chimes don't ring when you touch the doorbell switch wires together, remove the chime cover and vacuum out accumulated crud. If the chime plungers are rusty or corroded, spray them with WD-40® or a similar lubricant and move them by hand to get them sliding freely. Rubbing screw terminals with steel wool may improve electrical contacts. But chances are the old unit is worn out and needs to be replaced. To find out, dial a multimeter AC setting that's close to the low-voltage rating on the chime unit, then touch the tester probes to the *trans* and *front* terminals and to the *trans* and *rear* terminals, as shown in the photo at right. If you get a reading close to the unit's rating but the chimes won't ring, the transformer is delivering power, but the chime unit is defective. Replace it.

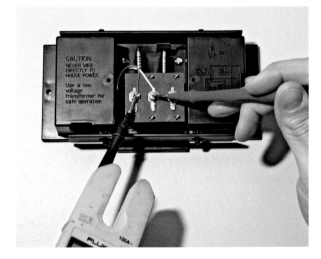

Test the chime terminals **using a multimeter.**

REPLACING A DOORBELL SWITCH

Replace a doorbell switch if it sticks, it is damaged, or continuity no longer exists between the contacts when the switch is depressed. You may also decide to upgrade simply because you want a newer style.

Unscrew the low-voltage wires and pull them out from the box for easy access. If they show corrosion, scrape or sand the wires lightly. Once the wires are clean and exposed, screw them to the terminals on the new doorbell switch. Then screw the new switch to the wall.

Choosing a new doorbell When replacing a doorbell switch/button pay close attention to the existing mounting holes. Purchase a replacement button that has the same hole pattern or a larger hole pattern. Installing a replacement button that is smaller than the original will require patching and painting of the original holes, which can be difficult and, depending on the mounting surface, may add to the amount of work necessary to complete the project. Installing a button with a hole pattern that does not match the existing button, but is close, can lead to problems when trying to install new screws too close to the existing screw holes.

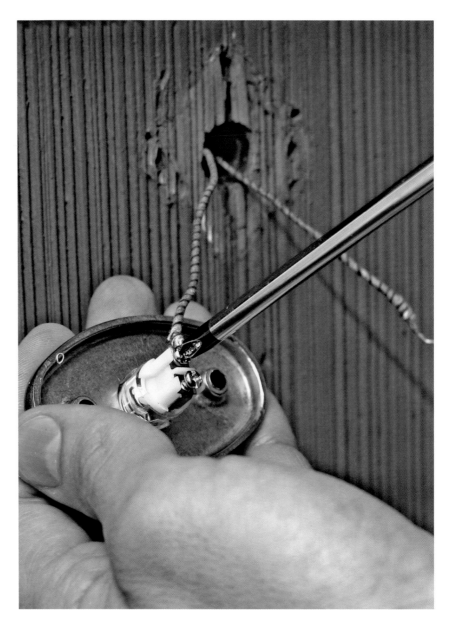

Screw the old wires **to the terminals on the new switch.**

INSTALLING A CHIME UNIT

Once you've determined that the old chime is defective, unscrew the low-voltage wires from its terminals and unscrew the mounting screws that secure the unit to the wall. As you carefully remove the old unit, hold on to the low-voltage wires so they don't fall into the wall. If the exposed wires are especially short, it's a good idea to tape them to the wall with duct tape.

After making sure the new unit has the same voltage rating the old unit, remove it from the packaging. Feed the low-voltage wires through the access hole in the new unit ❶. Level the unit's housing and then screw it to the wall. Because chime units are lightweight and installed in out-of-the-way places, it's seldom necessary to screw the chimes to the framing. It's usually sufficient simply to screw them to drywall or plaster ❷.

If the low-voltage wires look corroded or cracked and there's enough extra wire, snip their bare ends and restrip them. Then use needle-nose pliers to loop the ends clockwise ❸. Place the looped ends on the unit's terminals, exerting a slight pressure on the wires to keep them from slipping off as you tighten the screw terminals ❹. After you tighten the screw terminals, press the doorbell switch to test the new chimes. If they work correctly, snap on the unit's protective cover ❺.

1 Feed the wires from the wall into the new unit.

2 Level and mount the chime unit on the wall.

3 Strip and loop the low-voltage wires.

4 Attach the wires to the screw terminals.

5 Snap the cover into place and turn on the power.

TRADE SECRET
If doorbell wires are defective, try twisting new wires to the old ones, taping the splice, and pulling the new wires through the walls. If you don't succeed, install a wireless doorbell and chime system instead.

INSTALLING A DOORBELL TRANSFORMER

A doorbell transformer can be attached to any junction box that has enough room to accommodate the additional wire splices and is close enough to the doorbell switch that running wires to it is practical. But because junction boxes house wires with 120v voltage, you need to shut off the power to the junction box, carefully remove its cover, and then test to be sure the power is off ❶.

Once you've confirmed that no power is present, use a screwdriver or sturdy pliers to remove a knockout from the junction box ❷. Feed the new transformer's wires through the knockout opening ❸. Tighten the transformer's mounting screw to draw the unit tight to the junction box and cover the knockout opening ❹. (There is no need for Romex cable clamps or the like.)

If there is a green grounding wire from the transformer, screw it to the metal junction box, using a green grounding screw. If the box is plastic, splice the transformer's ground wire to the ground wire of the supply cable.

Strip the ends of the transformer wires and use twist-on wire connectors to splice them to the 120v supply wires. Splice like to like: Splice the hot (red or black) transformer wire to the hot supply wire; the neutral (white) transformer wire to the neutral supply wire ❺.

Once you have spliced the wires, tuck them into the junction box and replace the cover (the base of a porcelain light fixture is shown in the photo). Then connect the low-voltage wires to the screw terminals on the transformer ❻. Turn on the power and test the system.

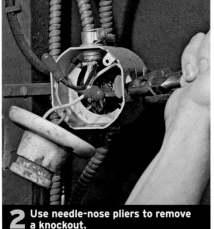

1 Turn off the power, then use an inductance tester to make sure it's off.

2 Use needle-nose pliers to remove a knockout.

3 Pull the wires from the new transformer through the knockout.

4 Mount the transformer to the box.

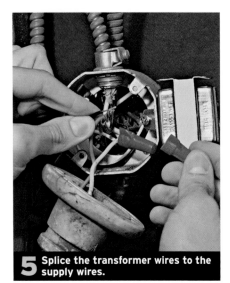

5 Splice the transformer wires to the supply wires.

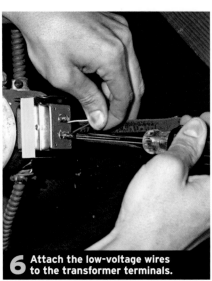

6 Attach the low-voltage wires to the transformer terminals.

TRADE SECRET

Before installing a new transformer, make sure it has the same voltage rating as the old one. The voltage rating should be stamped somewhere on the transformer housing.

TESTING AN OLD THERMOSTAT

If your thermostat is a low-voltage unit, it gets its reduced voltage from a transformer. If the thermostat doesn't turn the furnace on and off, remove its cover to see if it's clogged with dust or if the wires are loose ❶. Remove dust using a small painter's brush and reattach loose wires.

If the thermostat still doesn't work, use a continuity tester or a multimeter set to ohms (resistance) to test the unit. Unscrew the low-voltage wire attached to the R terminal and touch the tester probes to the R and W terminals. As you hold the probes to the terminals, flip or turn the setting lever from one side to the other ❷.

If the continuity tester lights up or the multimeter shows any reading at all, the thermostat works and the problem is the transformer. If there's no light or no meter reading, the thermostat is defective. Disconnect the remaining wire(s) and remove the unit ❸.

1 Remove the cover of the existing thermostat.

2 Test the terminals using a multimeter while adjusting the setting lever.

3 If defective, remove the existing unit.

> ## ⚠ WARNING
> Thermostats are specialized switches that open or close a circuit in response to temperature changes. Although most thermostats are low-voltage units that are safe to handle, your unit may be unsafe to handle if it mounts to an electrical box and is connected to house wiring. Shut off the power, test to be sure it's off, disconnect the thermostat, and have a heating professional assess it.

INSTALLING A PROGRAMMABLE THERMOSTAT

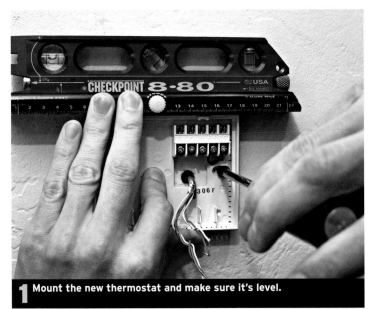

1 Mount the new thermostat and make sure it's level.

2 Attach the existing wires to the designated terminals on the thermostat.

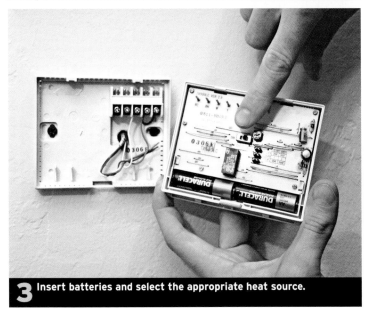

3 Insert batteries and select the appropriate heat source.

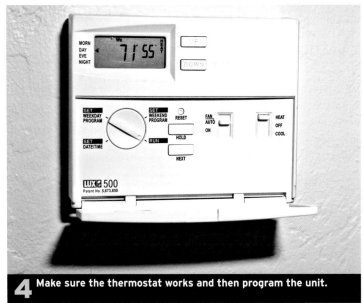

4 Make sure the thermostat works and then program the unit.

Though most thermostats are installed in the same way, be sure to follow the installation instructions that come with your programmable unit. Most thermostats require leveling to function properly **❶**. Once the unit is on the wall, attach the existing low-voltage wires to the designated terminals on the new unit **❷**. Either insert the stripped wire ends into push-in clamps or loop them clockwise around screw terminals. Many modern units have batteries that need to be changed periodically. You may also need to set a switch on the back of the thermostat plate to select the type of heat (electric, oil, or gas) **❸**. Replace the thermostat's cover, consult the instructions, and program the unit **❹**.

TRADE SECRET

If the thermostat works but its temperature settings are too high or low, it may need to be plumbed or leveled. Remove its cover and look for a pair of plumb or leveling marks on the base.

REPLACING A FURNACE TRANSFORMER

If your thermostat isn't getting power, the furnace transformer may be defective. If your furnace is relatively new, the transformer will be inside the furnace housing–in which case, call a HVAC (heating, cooling, air-conditioning) specialist to assess and replace it. New HVAC units are sensitive and complicated. However, if the transformer is mounted to a junction box on the outside of the housing, it's relatively easy to replace.

Set your multimeter to a low-voltage AC setting and touch its probes to the two terminals of the transformer ❶. If there's no reading, replace the transformer. Begin by shutting off the power and then using a voltage tester to be sure it's off. Then remove the old transformer and unscrew the cover on the junction box. (In the example shown here, the transformer and box cover are an integral unit.) After removing the cover, test the exposed house wires again to be sure the power's off.

Use a screwdriver or a sturdy pair of pliers to remove a knockout on the side of the box and insert the wire leads of the new transformer into the opening ❷. Tighten the diagonal screw that mounts the transformer to the junction box, then screw the transformer's green grounding wire to the metal box using a green grounding screw. If the box is plastic, splice the transformer's ground wire to the ground wire of the supply cable. Use wire connectors to splice like wire groups: connect the transformer's hot lead (black) to the hot supply wire and the neutral lead (white) to the neutral supply wire ❸.

After making these connections–but before covering the junction box–turn the power back on. Touch the probes on the multimeter to the transformer terminals to make sure the new unit functions normally ❹. Then turn the power off again. Snip, strip, and loop the existing low-voltage wires as needed and attach them to the transformer terminals ❺. Finally, gently push the wire splices back into the junction box and screw the cover in place ❻.

⚠ WARNING
Make sure the new transformer is properly rated for the furnace. Check the voltage rating stamped on the old transformer or look inside the cover of the furnace, which will list the low-voltage requirement for the transformer.

1 Touch the multimeter probes to the transformer terminals to test the unit.

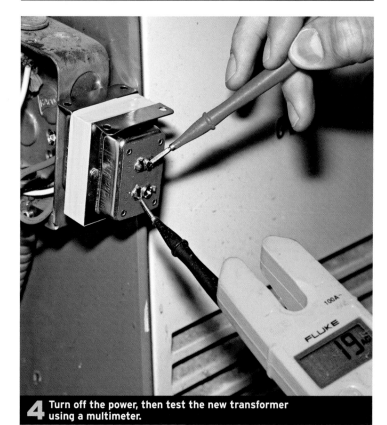

4 Turn off the power, then test the new transformer using a multimeter.

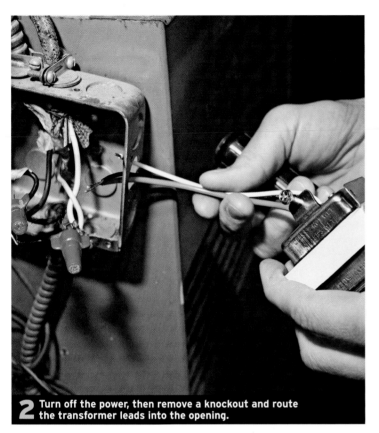

2 Turn off the power, then remove a knockout and route the transformer leads into the opening.

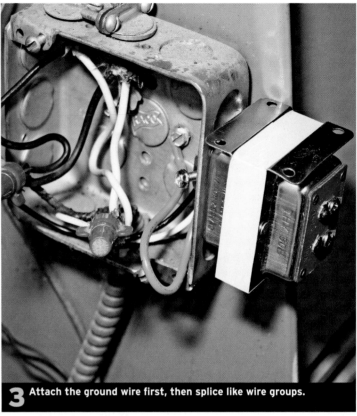

3 Attach the ground wire first, then splice like wire groups.

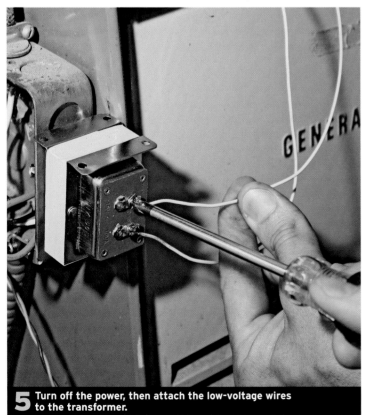

5 Turn off the power, then attach the low-voltage wires to the transformer.

6 Tuck the wires back inside the junction box and install the cover.

REPLACING A FURNACE TRANSFORMER | **119**

MULTIMEDIA

MULTIMEDIA ENCOMPASSES various types of video, sound, voice, and data signals, and have multiple format options. Wiring multimedia outlets and distribution panels is somewhat like assembling a computer, though prefabricated modules and interchangeable parts make home wiring fairly straightforward.

The tools and methods used to install cables are similar to those used to install basic residential electrical circuits. Whether you're adding a phone jack to existing service or installing a media distribution panel, you'll need only a few specialized tools, such as a long, flexible drill bit if you'll be running wires through walls. Multimedia connections are extremely precise, so it's important to follow installation instructions exactly to ensure strong signals throughout your sound, video, phone, and data networks.

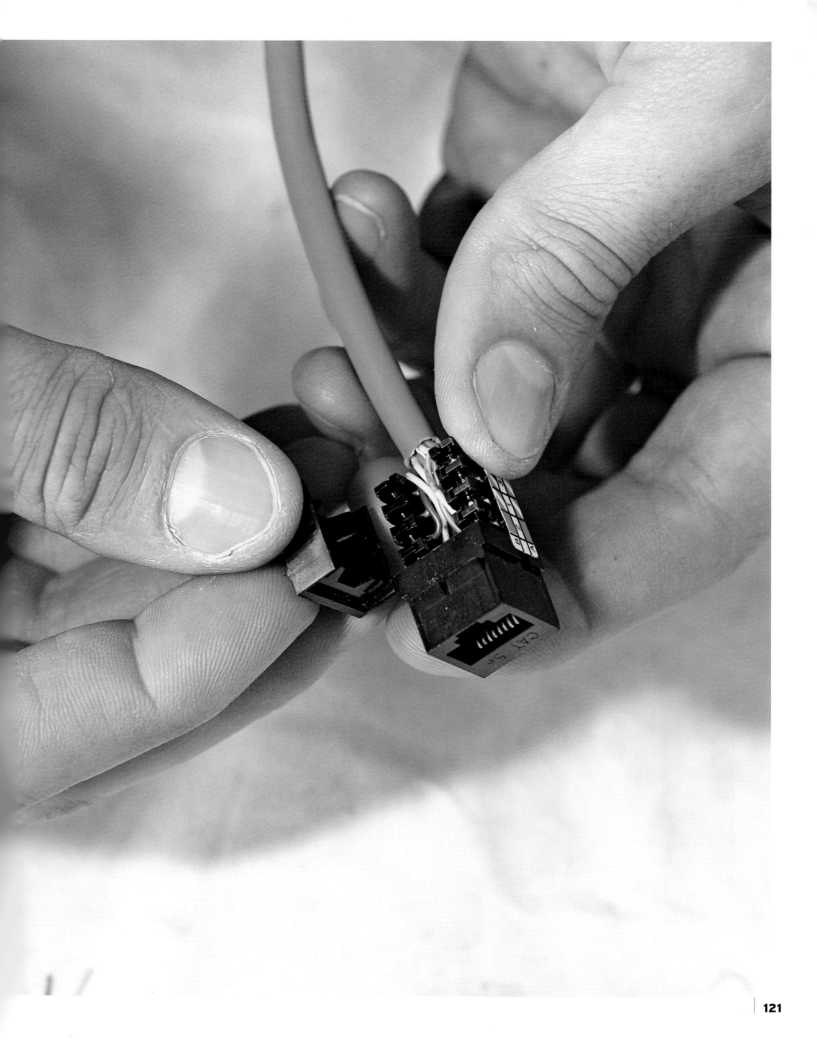

MULTIMEDIA CABLE & CONNECTORS

Here's an overview of the cables and connectors you'll need to wire your house for sound, video, data, and telephone service:

Coaxial cable includes double-shielded RG6 and RG59 cable and single copper-shielded RG59. RG6 cable has a slightly thicker wire gauge than does GR59. Used for video distribution (cable TV), coaxial cable has been around for years. Video cable terminates in an F-connector such as the watertight RG6 connector shown in the photo at right. To simplify life and forestall callbacks, many professional electricians install watertight F-connectors inside and out.

Data cable can carry data or phone signals; it is typically 24-guage, solid-wire unshielded twisted pair (UTP) cable. Although Category (Cat) 5e cable meets present-day standards for data transmission, Cat 6 wire is likely to supplant it. In general, the higher cable numbers denote faster data-transmission capabilities. Data cable contains four pairs of wires, thus RJ-45 data jacks contain eight pins.

Telephone cable is usually Cat 3 cable, which is not twisted; its solid wire is considerably thinner than Cat 5e cable. RJ-11 phone jacks contain six pins. Although data cable can carry phone or data signals, Cat 3 phone cable is not rated for high-speed data transmission.

Audio (speaker) cable is usually 18-gauge to 12-gauge finely stranded wire. Speaker wire terminations vary from bare wires compressed between stacked washers to screw-on or crimped jacks that plug into speaker ports. Plug-in jacks are color striped to match speaker polarity: red-striped jacks for positive terminals and black-striped jacks for negative terminals.

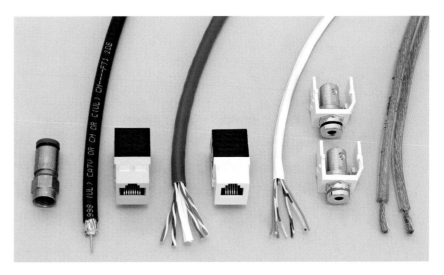

Each multimedia connector is to the left of the cable it terminates. **From left: RG6 F-connector, dual-shielded GR6 coaxial cable; RJ-45 (eight-pin) jack, Cat 6 UTP data cable; RJ-11 (six-pin) jack, Cat 3 phone cable; two audio jacks (sometimes called** *banana jacks***), 14-gauge low-loss audio cable.**

Signal strength: coming in loud and clear

Although solid connections are as important to multimedia as they are to all electrical systems, signal strength—not voltage—is the objective when connecting data, sound, video, and phone components. In fact, most multimedia input is not impelled by AC, as is house wiring. Rather, video and phone signals are generated by cable or phone companies. Computers and routers amplify data signals, and stereo amplifiers boost sounds signals; but although those devices run on house current, the signals themselves are not AC. Thus the cables that carry multimedia signals are dramatically different from, say, Romex cables, and require different connecting devices and a few specialized tools.

TRADE SECRET

Pros use watertight connectors on all coaxial cable connections, even those installed indoors. Watertight connectors such as the RG6s shown in the photo above don't cost much more and they always provide a solid connection.

STRIPPING CABLE

Electricians generally favor one type of stripper and use it to strip everything, thus reducing the number of tools in their belts.

Splicing scissors can trim tiny wires, but they can also score sheathing: Hold the cable in one hand, and, with the other, hold a scissors blade perpendicular to the cable and rotate it around the cable **A**. It's not necessary to cut through the sheathing. Once scored, the sheathing will strip off when you pull on the scored sections **B**. In fact, merely scoring the sheathing is less likely to damage individual wire insulation.

The cable scorer shown in photo **C** is faster than using splicing scissors, but its razor blade is so sharp that it can easily nick wire insulation if you're not careful. Wire strippers are the most reliable way to strip insulation off individual wires because you can choose a stripper setting that matches the wire gauge **D**. Finally, whether stripping wire, drilling holes, or doing any other wiring task, wear eye protection and sturdy gloves.

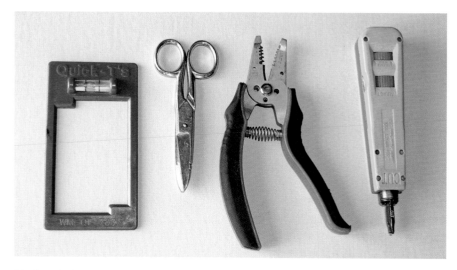

Multimedia installation tools. From left: low-voltage cut-in ring template, splicing scissors, wire strippers, and punch-down tool.

TWISTED WIRES

Remove the sheathing from Cat 5e or Cat 6 cable, and you'll discover twisted wires within. Twisting wires reduces the occurrence of *cross-talk*, in which electro-magnetic signals jump from wire to wire. Although it is necessary to straighten some data cable wires to attach them to some jacks, the twisting must be maintained to within 1/4 in. of the termination on the jack.

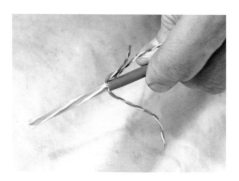

Cat 5e and 6 cables are examples of UTP cable, which is twisted to prevent signals from jumping between wires.

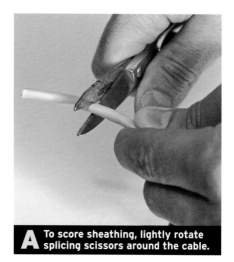

A To score sheathing, lightly rotate splicing scissors around the cable.

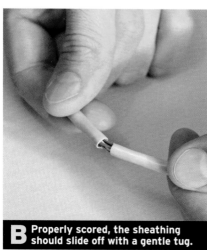

B Properly scored, the sheathing should slide off with a gentle tug.

C Lightly twirl a cable scoring tool around the sheathing.

D Wire strippers have settings that match wire gauges.

EXTENDING A PHONE LINE

Thanks to cell phones and remotes, we're no longer tied to the room where the phone line ends. But for clear, reliable service you can't beat a hard-wired phone plugged into a nearby jack. Happily, almost anyone can run an extension from an existing jack, thereby saving a hefty installation fee from the phone company.

The only tricky part of the job is running the cable to the new jack. You can tuck it under carpets, tack it atop baseboards or run it around door jambs, but it won't look great. For the cleanest job, route that extension line into the attic or basement or—as shown here—drill through the wall and run it outside.

Before you start, find the shortest route between the existing jack and the new one. Measure that route carefully and add enough extra cable for drip loops and at least 1 ft. extra on each end for the thickness of walls and for stripping and connecting to the jacks. Buy Cat 3 cable, which contains four pairs of solid-core, 24-gauge wire. Because phone lines are low voltage, they're safe to handle. You can do the job with common tools.

Connecting to an existing jack

Start by unscrewing the cover on the jack and the mounting screws that hold the jack to the wall. Gently pull the jack out from the wall, being carefully not to pull loose the wires attached to the jack terminals ❶.

If the jack is surface mounted, drill a hole for the extension cable that will be covered by the jack. If the jack is flush mounted, drill anywhere inside the ring, because the hole will be covered by the jack cover. Drill at a downward angle ❷ so that exterior water will tend to run out of the

1 Gently pull the jack out from the wall.

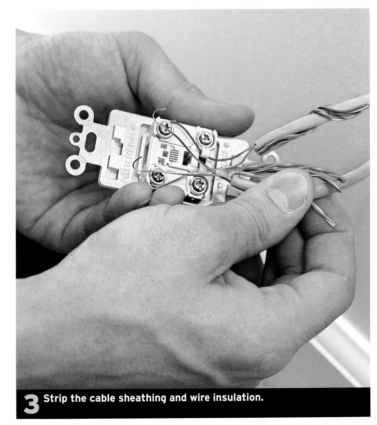

3 Strip the cable sheathing and wire insulation.

TRADE SECRET

Many electricians install Cat 5e cable for both data and phone lines to allow future expansion or modification. That is, Cat 3 is fine for phone service but inadequate for data, whereas Cat 5e can carry both signals.

hole. Look into the outlet opening before drilling to avoid electrical cables in the wall. Use a ¼-in. extension bit in a cordless drill for the job.

After drilling through the wall, use duct tape to attach a piece of string to the bit and pull the bit back through the wall. Then tie the new Cat 3 phone cable to the string and pull it into the hole behind the jack. Once you've pulled the new cable to the existing jack, strip about 2 in. of the cable sheathing and separate the cable wires into pairs ❸.

Using splicing scissors or a wire stripper, strip about ½ in. of insulation from a wire pair (for example, a blue and a blue-white wire), loop the bare wire ends clockwise, and attach one wire to each of the two jack terminals that are presently wired ❹. Typically, terminals have stacked washers that hold several wires.

2 Drill downward to prevent water from traveling to the interior.

Wiring techniques

When connecting to an existing phone line, avoid bending existing wires repeatedly because they can become weak and break off. Screw terminals come with multiple washers and are intended to have one wire under each washer. It is not recommended to twist wires together and install them under one washer. Finally, always wrap wires around screw terminals in a clockwise direction.

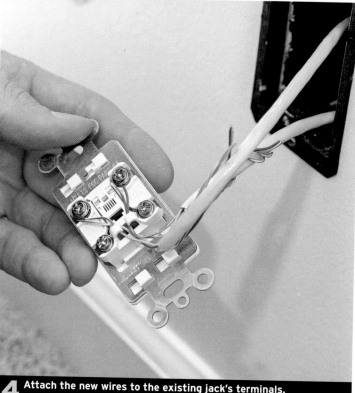

4 Attach the new wires to the existing jack's terminals.

TYING INTO A NEW JACK

Before pushing the cables into the wall and remounting the existing jack, staple the cable so it can't move and stress electrical connections. If the cable runs outside, loop it downward so water will drip off, and staple it with insulated cable staples ❶. Then fill the hole in the siding with siliconized latex caulk.

Locate the new jack and drill through the wall to bring cable to the location ❷. As described earlier, tape a string to the end of the drill bit before withdrawing it, then tie the new cable to the string. Once you've pulled the new cable into the room, remove its sheathing, strip insulation from a pair of wire ends, feed the cable through the new jack, and screw the jack to the wall ❸.

Loop the bare wire ends clockwise, insert them between the stacked washers on the jack, and screw them tight ❹. Finally, tug gently on the wires to be sure they're well attached, tuck the wires neatly out of the way, and snap on the jack cover ❺.

> ### ⓘ TRADE SECRET
> When running low-voltage wire or multimedia cable, staple it at least every foot to prevent sagging, strains on connections, and an unsightly appearance.

1 Form a drip loop, then staple the cable to the wall.

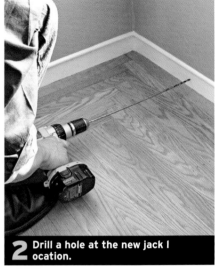

2 Drill a hole at the new jack location.

3 Mount the new jack in the opening.

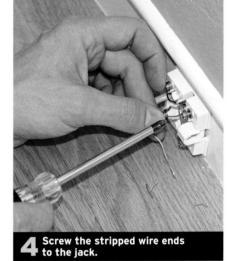

4 Screw the stripped wire ends to the jack.

5 Snap on the jack cover and double-check to make sure it works.

> ### ⓘ TRADE SECRET
> If it's necessary to drill through an exterior wall to bring lo-vo wiring or media cables into a room, drill downward to keep water from running into the walls. After inserting the wire or cable, fill the hole with siliconized latex caulk which, unlike pure silicon, is paintable.

MOUNTING A MEDIA DISTRIBUTION PANEL

A distribution panel is the heart of a home multimedia network because it receives incoming signals from phone, cable, and data companies and distributes them throughout a house. Panels can also distribute the signals from a stereo or sound system to speakers in many rooms. Many panel makers offer solutions with prewired modules, so a homeowner can configure a multimedia network today that can be easily changed tomorrow. To successfully install your system, closely follow the instructions provided, using the recommended tools.

As with other electrical installations, connect the power last, after wiring the panel and running cable to the outlets. Wear safety glasses at all times and work gloves to protect your hands. Panel edges can slice your hands in a flash; metal shards dislodged by hammers or drills are also dangerous. Finally, sturdy gloves enable you to grip and pull cables more easily.

Mounting the panel

Locate the distribution panel centrally so you'll have shorter cable runs: None should be longer than 300 ft. from the panel to an outlet. To minimize electrical interference, locate the multimedia panel away from the service entrance panel or a subpanel.

Ideally, place the distribution panel on an interior wall to minimize temperature fluctuations—never locate it in an attic or unheated garage. You'll build in future flexibility if you run Cat 5e or Cat 6 cable for all data and telephone lines.

Install the rough-in box for the multimedia distribution center by screwing it to adjacent studs **1**. If you've already pulled most of cables to feed the panel, as shown here, tack them to one side so they'll be out of the way. Slide the panel in or out so its edges will be flush with the drywall **2**.

>> >> >>

1 Screw the panel to the studs.

2 Set the edges of the panel flush with the drywall.

> **TRADE SECRET**
> When mounting panels, install screws at the top loosely to hold the panel in place. Set and secure the bottom, then reset the top screws as needed.

MOUNTING A MEDIA DISTRIBUTION PANEL (CONTINUED)

Remove the knockout(s) for the power supply ❸. If your panel will house only telephone and cable TV, you may not need a power supply, but routers require line voltage, so installing a power supply module will provide future flexibility. Drop the power-supply unit into the opening, run Romex cable to it, and screw the module to the panel housing ❹.

Install the final locknut on the utility conduit ❺. The conduit may run outside to the utility pole or to an underground pull box. Depending on your utility's main point of entry (MPOE) or demarcation, you may need an exterior box to which the utility will run the service. Or the utility may install the MPOE inside, in the distribution panel.

3 Remove a power-supply knockout.

4 Attach a power-supply module to the panel.

> ⚠️ **WARNING**
> Plan your interior layout before removing panel knockouts. That way you know where and how you want your cables to come in and you don't have to go back to reroute cables and seal unused knockouts.

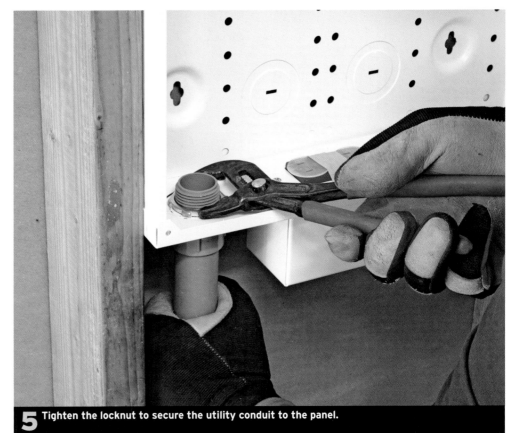

5 Tighten the locknut to secure the utility conduit to the panel.

A TYPICAL MULTIMEDIA PANEL

From this one multimedia panel (its outgoing low-voltage wiring has been completed), cables distribute video, sound, telephone, and data signals throughout the house. In this panel, the incoming utility and cable services are not yet installed—they will be pulled in through the conduit at the lower left.

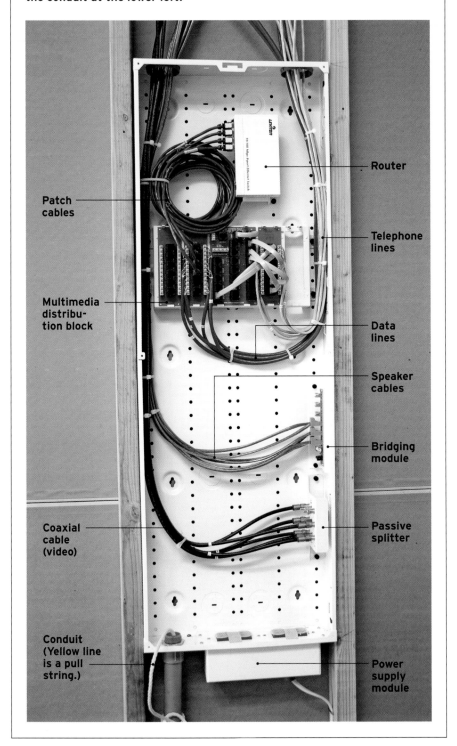

Patch cables

Multimedia distribution block

Coaxial cable (video)

Conduit (Yellow line is a pull string.)

Router

Telephone lines

Data lines

Speaker cables

Bridging module

Passive splitter

Power supply module

Panel basics

A *router* distributes digital subscriber line (DSL) or broadband signals from a *modem* (not shown) throughout the house so you can have several people accessing the Internet at one time. *Patch cords* make connections from the router to jacks on the *data/phone distribution* block. The *multimedia distribution* block is a prebuilt unit: One side has banks of jacks and punch-down blocks that distribute data signals to outlets; its other side distributes telephone lines. A *passive splitter* distributes video signals from a cable company throughout the house; an *amplified splitter* boosts the video signal before distributing it. A *bridging module* distributes sound signals to speakers around the house. The *power supply module* is fed with 120v house current; it has receptacles that router or amplified splitter cords (not shown) can plug into.

RUNNING CABLES TO A MEDIA DISTRIBUTION PANEL

There's no one right way to run cable to a multimedia distribution panel. In new construction, electricians typically start from the distribution panel and pull cable outward. But remodel wiring is rarely predictable because you can't see obstructions hidden in walls and floors.

Before pulling cable into the panel, snap plastic bushings into the panel knockouts so sharp edges won't chafe the cable sheathing ❶. As you pull cable into the panel, loop it gently so that you don't crimp it ❷. If you have a lot of cables entering a panel, roughly divide them between two or more knockouts so the box will look neater.

Install prewired data and phone distribution boards and other modules. In photo ❸, the two columns on the left are jacks with 110-type punchdowns to which you'll attach data cables. Incoming phone lines terminate in the center of the module, which are then distributed to phone jacks throughout the house, permitting multiple lines at different locations.

As you route cable through the panel, tie-wrap cable bundles to free up workspace and enable you to see connections easily ❹. Labeling both ends of every cable is also essential, so you'll know which cable is which, should you need to repair or modify the multimedia wiring ❺. Finally, create a numbered house map to show the cable locations.

1 Use snap-in bushings to protect cables.

2 Loop the cable to avoid crimping it.

3 Install modules in the panel.

4 Bundle the cables to conserve space.

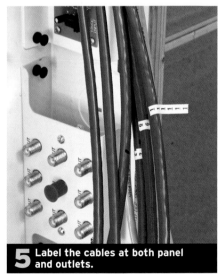

5 Label the cables at both panel and outlets.

CONNECTING CABLES AT THE PANEL

Because different types of cables converge at the distribution panel, there are, naturally, different connectors used to attach each one.

Speaker wires are typically stranded. Strip the ends as indicated by the panel maker—¼ in. is typical—insert them into the connectors' screw-down terminals and tighten them securely **1**. Some panels will terminate speaker wires in a banana jack, as shown in the photo on p. 122, here, a plastic jack snaps to a bridging module **2**.

Using a combination stripping and crimping tool, strip the coaxial cable that carries data or phone signals and crimp an F-connector into the cable **3**. Plug coaxial cables into the passive splitter and tighten them snug **4**. (*Note: The capped blue terminal is for input from the cable company.*)

Strip approximately 3 in. of sheathing from the ends of Cat 5e and Cat 6 cable (data and phone); separate the stranded wire pairs; and, using a punchdown tool, press the individual wires onto the insulation displacement connector (IDC) prongs on the data and phone board **5**. The punchdown tool also removes excess wire. The blue cables in the foreground are data cables; the beige cables are phone lines **6**. Finally, plug in patch cords that connect router input to jacks on the data and phone board **7**.

1 Screw down the speaker wires in the terminal.

2 Insert the speaker jack into the bridging module.

3 Strip and crimp the coaxial cable.

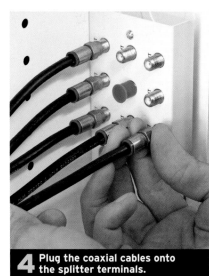

4 Plug the coaxial cables onto the splitter terminals.

5 Install the data and phone wires.

6 Once connected, group and strap the cables to keep the panel neat.

7 Use patch cords to connect the router to the data.

LOCATING THE MULTIMEDIA OUTLET

1 Align the new outlet with a nearby receptacle.

2 Level the cut-in ring and align it with nearby receptacles.

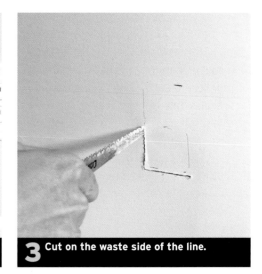

3 Cut on the waste side of the line.

4 Use a flexibit to drill access holes for the cables.

5 Hook the kellum to the bit and pull the cable to the outlet.

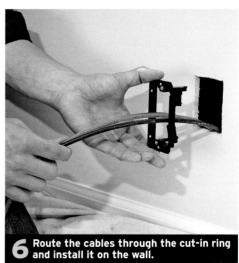

6 Route the cables through the cut-in ring and install it on the wall.

In general, electricians run one data, one phone, and one video cable to each outlet. In the example here, however, specs dictated two data lines (green and blue cables) and a coaxial cable for video.

When locating the cut-in ring for a new outlet set alongside an existing receptacle, always measure from the center of the existing box, so the center screws on the cover plates will line up **1**. The two types of cover plates may be different sizes, so lining up their tops or bottoms won't look good. Level the cut-in ring and trace its outline on the wall. (The template shown in the top photo

on p. 123 is another option.) Don't try to eyeball level: If you're even slightly off, the outlet will look cockeyed **2**.

Use a drywall saw (also called a jab saw) to cut out the opening. To start the cut accurately, strike the heel of your hand against the saw handle. To avoid cutting too large a hole, cut on the waste side of the outline **3**.

To bring cables from the distribution panel, drill holes into the wall plate using a flexibit **4**. Wear gloves or use a drilling guide to protect your hands while guiding the bit. Flexibits have a hole in the flute (cutting edge). After the bit emerges through

the wall plate, have someone tape the cables together, slide a swivel kellum over the taped wires, and hook the kellum to the bit **5**. The kellum swivels, so the wires won't twist up as you reverse the drill and pull them back up through the hole. Install the cut-in ring to provide a mounting surface for the outlet plate **6**.

➤ For more on running wires, see p. 199.

INSTALLING CABLE CONNECTORS

In the next few pages, we'll discuss the installation of several different types of cable connectors: F-connectors used on coaxial cable (video) as well as two popular systems (from Leviton® and Panduit®) that connect to Cat 3 cable (phone), Cat 5em and Cat 6 cables (telephone or data). Each connector maker specifies tools and methods of assembly, so follow its installation instructions closely.

Use a combination stripper-crimper to strip insulation off the end of the coaxial cable. Stick the cable in the end of the tool, spin it, and peel off both types of insulation from the outer sheathing ❶. The tool leaves about ½ in. of bare copper and ¼ in. of white insulation with the shielding on it. Slide an F-connector over the stripped end of the coaxial ❷. Place an F-connector in the crimping bay, insert the stripped cable, and squeeze the tool to crimp the connector tight to the cable ❸. (*Note:* The F-connector shown here is a watertight variety.)

TRADE SECRET

It's not necessary to house multimedia connections in covered junction boxes (required for high-voltage splices). Install flush-mounted cut-in rings to provide a secure device to which you can attach outlet plates and insert jacks. Whether you're mounting doorbell chimes or a multimedia wall outlet, take the time to level and plumb the device and, if there's another outlet nearby, to align the height of the new outlet.

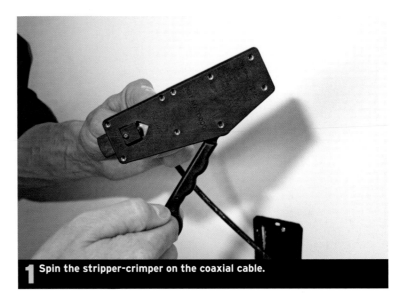

1 Spin the stripper-crimper on the coaxial cable.

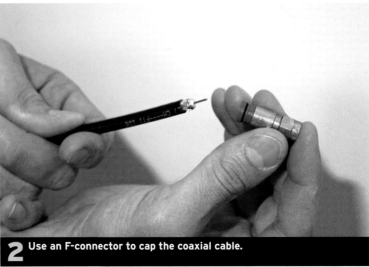

2 Use an F-connector to cap the coaxial cable.

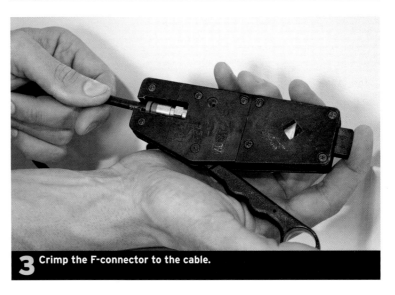

3 Crimp the F-connector to the cable.

CONNECTING PANDUIT JACKS

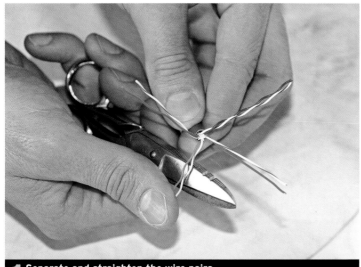

1 Separate and straighten the wire pairs.

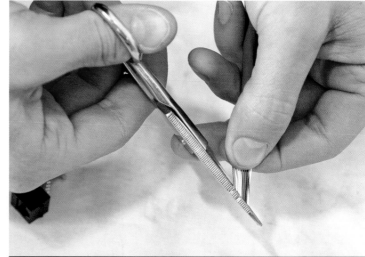

2 Cut the wire ends diagonally.

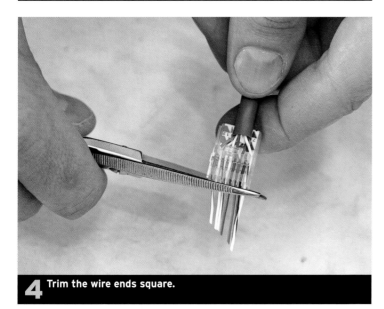

4 Trim the wire ends square.

5 Pair a Panduit jack housing with a wired cap.

Connecting phone or data cable jacks to a multimedia outlet requires a few steps. Start by reviewing cable stripping (see p. 123). Cat 5e and Cat 6 cable are UTP cable, so after stripping about 3 in. of sheathing, separate the wire pairs before attaching them to a jack. If you use Panduit jacks (often referred to as *Pan jacks*), you'll also need to unwind (untwist) the wires. (Cat 3 phone wires are straight and so do not need untwisting.)

Use stripping scissors to unwind the individual cable wires before attaching them to a Pan jack ❶. Using light pressure, pull the wires across a scissor blade to straighten them. Next, slide the wires into a plastic cap, which is color coded to indicate where the wires go. Flatten the wires and cut their ends diagonally so they'll slide easily into the cap ❷. Push the wires all the way into the cap ❸, then trim the wire ends straight across ❹.

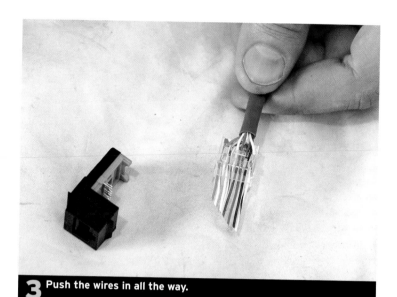

3 Push the wires in all the way.

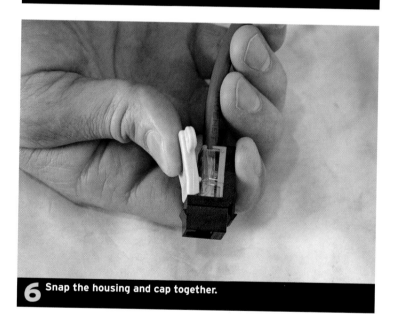

6 Snap the housing and cap together.

The wired cap snaps into a jack housing, forcing the wires into V-shaped IDC prongs that slice the wire insulation to make the electrical contact **5**. Place the wired cap onto the jack housing and use the small plastic lever provided to snap the assembly together **6**. Panduit jacks are reliable because they make secure connections.

INSTALLING LEVITON JACKS

The Leviton jack system uses the 110 punchdown tool required to punch down wires in the distribution panel (see photo 5 on p. 132). But the system takes practice to avoid loose wires, so amateur electricians might get more predictable results using Panduit jacks.

In the Leviton system, there's no need to untwist wires. Separate the wire pairs and punch them directly into the IDC prongs built into the jack. Use the plastic disc provided to back the jack as you punch down **1**. The punchdown tool comes with 110 and 66 blade sizes; each blade has a side that punches the wire down and a side that cuts off excess wire. For best results, work from back to front to avoid disturbing wires that are already down. Once you've connected the wires, snap on the jack's trim cap, which keeps wires in place and relieves strain on the connection **2**.

1 Punch down the wires to the Leviton jacks.

2 Snap the trim cap into place.

ATTACHING JACKS TO AN OUTLET PLATE

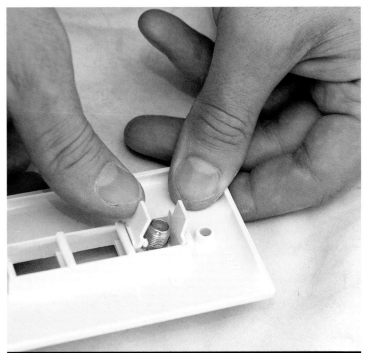

1 Snap on the screw-in coupling for coaxial cable

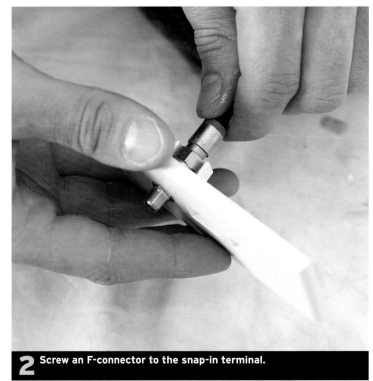

2 Screw an F-connector to the snap-in terminal.

In the project shown here, we're installing both Leviton and the Panduit jacks to show that despite variations in jack wiring, both snap into the most common type of face plate, the keystone style.

To install coaxial terminals, snap the threaded coupling into the face plate **1**. The coaxial cable's F-connector screws to the coupling, creating a clean termination instead of having cable hanging out of the wall **2**. Snap a RJ-45 (eight-pin) Leviton jack into the face plate **3**. Note the color coding on the side of the jack to indicate the order of wires you punched down earlier. Finally, snap a RJ-45 Panduit jack into the remaining port in the keystone plate **4**.

Carefully feed the cable into the wall, hold the face plate flush to the wall, and screw it to the cut-in ring **5**. The outlet shown uses phone and data jacks that are different colors so that users can quickly differentiate which jack is which. This differentiation is not an issue in residences, but it's imperative in business installations.

Faceplate options

There are many types of faceplates on the market for mounting telephone, data, coaxial, and specialty jacks. They come designed to match the adjacent electrical outlet (for example, duplex or Decora® style) and in various colors. Although it is beyond the scope of this book, there are also faceplates and adapters for an array of applications such as video (HDMI, S-video, component video, VGA), sound (binding posts, banana), and fiberoptics.

3 Snap in the Leviton data jack.

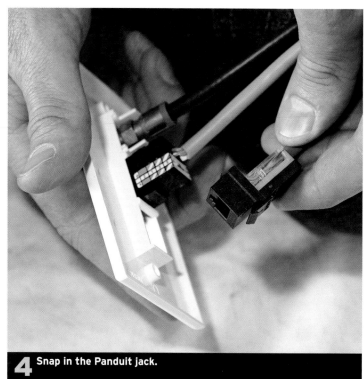

4 Snap in the Panduit jack.

5 Screw the face plate into place.

FANS

RESIDENTIAL FANS CAN BE divided into two groups. The first group might be better called exhaust or vent fans, because their primary function is to remove moist or odor-laden air from living spaces, route it through ducts, and expel it outside. Commonly installed bath fans and range hoods both fall into this category. By removing excessive interior moisture, these fans help forestall mold and other unhealthy conditions.

The second group, primarily ceiling fans, circulate still, summer air and thus increase comfort by evaporating moisture on your skin. In cold climates, these fans push warm air down from the ceiling where it collects, so it can once again warm the bodies below. Whole-house fans, installed in attics in warm regions, also circulate hot air, but because they typically push it out roof or gable-end vents, whole-house fans are more like exhaust fans without ducts.

BEFORE YOU BEGIN

There are two important tips to note before wiring any type of fan. First, check out the wiring schematic that comes with the unit **A**. In most cases, the schematic is affixed either to the fan housing or to the backside of a cover. Or it may be included in the installation instructions. You'll also find essential information such as the fan's rating, expressed in amperes.

⚠ WARNING

When installing or replacing a fan of any type, always turn off the electrical power to that location. Then test with a voltage tester to be sure the power is off.

Second, most units have an integral junction box, like the one seen in photo **B**. The junction box contains wire leads that you'll splice to the incoming house wiring. All metal boxes—including fan housings—must be grounded, so if the fan maker doesn't include a grounding pigtail to the housing, add one.

ⓘ TRADE SECRET

Never assume that all the parts you need for assembly and installation are in the box. Check the contents before you start by comparing the parts in the box to the inventory list included in the owner's manual. Smaller parts, such as screws, are typically packaged in clear plastic so you can count them without having to open the packet—a good idea because they're easy to lose.

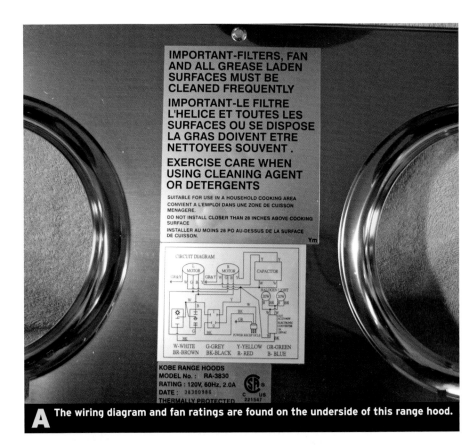

A The wiring diagram and fan ratings are found on the underside of this range hood.

B An integral junction box is in the housing of this bathroom vent fan. The second green wire grounds the metal housing.

CHOOSING A BATHROOM FAN

Bathroom fans are increasingly powerful, quiet, and available with numerous bells and whistles. When considering all the extra features, however, remember that the primary function of a bathroom fan is to remove moisture.

There are complex formulas for sizing bath fans, but a good rule of thumb is 1 cfm (cubic foot per minute) per square foot for bathrooms 100 sq. ft. or smaller. For bathrooms larger than 100 sq. ft., allot 50 cfm for each fixture (toilet, lavatory, shower) and 100 cfm for hot tubs.

Also, get a quiet fan: a rating of 3 sones to 4 sones is tolerable, 1 sone is very quiet. And remote inline fans, typically installed in attics some distance above bathrooms, are quieter still. Consequently, you can install a larger fan inline and still have a quiet bathroom.

Next, consider switches. Fans usually need to continue venting after you leave the shower or use the toilet, so get an electronic switch with an integral timer so that the fan continues running after the light is turned off. If there's no integral timer on the switch, you can run the light and the fan on separate switches. You can also connect the fan to a *humidistat*, which is a moisture sensor that turns the fan off once a preset moisture level is attained.

WARNING

Never install a rheostat—commonly called a dimmer switch—to control an exhaust fan. Fan motors are designed to run on a fixed voltage, and if you reduce the voltage by using a rheostat, you may burn out the fan motor.

FAN COVERS

Most exhaust fans come with plastic covers, but they can be replaced with metal plates. Here, an antique wall register was plated to match the other fixtures in the room.

REMOTE INLINE FANS

Even a well-made bathroom fan will be relatively noisy if the motor is 2 ft. from your head in the bathroom ceiling. But if you install the fan some distance from the bathroom, you'll reduce the noise considerably. That remote location may mean that you have longer duct and wiring runs, but routing them is rarely a problem. In fact, with a large-enough fan motor and a duct Y-connector, you can vent two bathrooms with one fan. Because longer duct runs can mean greater air resistance, consider installing rigid-metal or polyvinyl chloride (PVC) ducts, whose smooth surfaces offer less resistance, rather than flexible metal ducts. Alternatively, you could oversize the fan slightly. Better fan makers, such as Fantech® and American Aldes™, offer acoustically insulated cases to deaden sound further.

Remote inline fans **require longer ducting, but they're quieter because they're located farther away from the bathroom.**

LOCATING A BATHROOM FAN

There are three primary considerations when locating a bath fan. First, place the fan where it can expel moisture effectively—ideally, near the shower, where most of the moisture is generated.

Second, locate the fan so that its duct run is as short as possible and minimizes cutouts in the blocking or framing members—especially rafters. Vent the ductwork from the fan out the roof or through a gable-end wall. Avoid sidewall vents because moist air expelled by them could be drawn up into the attic by soffit vents in roof overhangs.

Third, locate roof vents away from problem areas such as skylight and valley flashing. Water usually dams up on the uphill side of a skylight, creating a leak-prone area that must be carefully flashed. Typically, skylight flashing consists of two pieces: a base flashing and a counterflashing that goes over it. Locate a roof vent near skylight flashing and you're inviting trouble. Valley flashing, on the other hand, may consist of a single broad piece of metal or elastomeric material folded up the middle. But because this flashing is located where roof planes converge, it channels an enormous amount of water during rainy seasons. So keep things simple: Locate roof vents away from obstructions in the roof or concentrations of water. Don't put a roof vent near an operable window, either.

A BATHROOM FAN

To keep moisture from leaking into the attic, apply silicon caulk between the fan box flanges and the ceiling drywall. Use metal foil tape to ensure airtight joints where ducting attaches to fan and vent takeoffs.

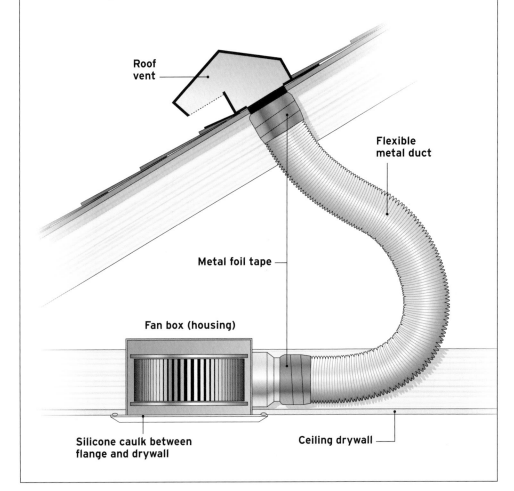

Roof vent

Flexible metal duct

Metal foil tape

Fan box (housing)

Silicone caulk between flange and drywall

Ceiling drywall

TRADE SECRET
Before you cut a hole in the ceiling, be sure there are no obstructions along the way. Tentatively locate the fan and use a 1/4 in. extension bit to drill along the proposed duct path. You'll also want to drill an exploratory hole up through the roofing.

INSTALLING THE ROOF VENT

1 Lift up on the shingle to remove the roofing nails.

2 Remove the shingles over the entire section.

3 With the vent in place, trim the shingles to fit the upper arc of the vent.

4 Use a utility knife to cut the shape of the vent hole through the roofing paper.

In the installation shown here, there was enough clearance around the exploratory hole and there happened to be a roofer on site, so the crew decided to install the roof vent first. (Roof vents vary; the model shown has a round stack and a weatherproof cap.)

Go onto the roof and find the ¼-in. exploratory hole drilled while locating the fan. The hole represents the center of the vent hole you'll need to cut.

Most ducting and roof-vent takeoffs are 4 in. diameter, so sketch that circle onto the roof. If the circle would cut into the tabs of any shingle—roughly the bottom half of a shingle strip—use a shingle ripper or a cat's paw to remove the nails holding those shingles in place before cutting the vent hole **❶**. Be gentle when removing shingles so you can reuse them **❷**.

Slide the upper flange of the roof vent under the shingle courses above and use a utility knife to trim its arc onto the shingles above **❸**, then cut out the vent's circle into any remaining shingles and the roofing paper **❹**. Next, use a

>> >> >>

INSTALLING THE ROOF VENT (CONTINUED)

5 Cut the roof sheathing to allow for the vent.

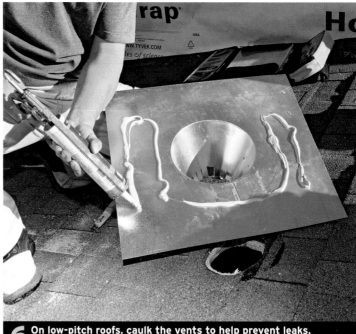

6 On low-pitch roofs, caulk the vents to help prevent leaks.

reciprocating saw to cut through the sheathing **5**.

If the roof pitch is 4:12 or greater, it usually isn't necessary to caulk the vent edges. Here, a 2:12 pitch required caulking to forestall leaks **6**. Carefully lift the shingle course above the vent and nail the two corners of the vent's upper flange into place **7**. Do not nail the lower corners of the vent: those nails would be exposed to weather and could leak. Slide in the surrounding shingles and renail them.

TRADE SECRET

Keeping fan-expelled moisture out of attics and wall cavities is crucial, and the only way to do so is to create airtight connections: Caulk the fan housing to the ceiling and seal each duct joint with aluminum foil tape, not fabric duct tape.

7 Nail only the top corners of the vent—shingles cover these nails.

MOUNTING THE FAN BOX

1 If needed, add blocking between the framing members.

2 When positioning the fan, a scrap of drywall acts as a stand-in for the finished ceiling.

3 Slide on the flex duct and tape it in place.

If bathroom framing is exposed, mounting the fan is pretty straight-forward. If you remove the fan assembly from the fan box, the box will be lighter and easier to hold in place one-handed while you use your other hand to screw the unit to the ceiling joists. Most fans have expandable brackets which extend between joists spaced 16 in. on-center (o.c.). But you should always screw at least one side of the fan box to a joist, to ensure that it's anchored securely. For ceiling joist (or rafter) spacing greater than 16 in. o.c., it's a good idea to add blocking **1**.

If the fan box flange mounts flush to the underside of the ceiling, use a piece of drywall scrap to gauge the depth of the unit relative to the finished ceiling **2**. Regardless of whether the box flange sits above or below the ceiling drywall, caulk the flange with polyurethane sealant to create an airtight seal between the two materials. If you removed the fan assembly earlier, reinstall it now.

Keep duct runs as short as possible to reduce air resistance. Slide the lower end of the flexible duct to the fan's exhaust port **3** and seal the joint with metal duct tape, then attach the other end to the roof vent takeoff. Or, if you haven't yet cut the hole in the roof, hold the free end of the duct to the underside of the roof sheathing and trace its outline onto the surface.

RETROFITTING A BATHROOM FAN

If you are remodeling or installing a bath fan and the finished ceiling is already in place, begin by creating a cardboard template of the fan housing. Mark the approximate position of the fan by driving a screw or nail through the ceiling and then go up into the space above the bathroom and find the marker. If there is an insulated attic above, take along a dustpan to shovel loose insulation out of the way and be sure to wear a dust mask and gloves. After you've located the marker, place the fan template next to the nearest joist and trace around it. Most fan boxes mount to ceiling joists. If the fan box has an adjustable mounting bar, you have more latitude in placing the fan. Use a jigsaw or reciprocating saw to cut out the opening. To keep the drywall cutout from falling to the floor below, take a piece of scrap wood slightly longer than the cutout and screw it to the drywall.

Before placing the fan box on top of the ceiling drywall, caulk around the perimeter of the opening to create an airtight bond to the box flange ❶. Once the box is wired and the ceiling has been painted, install the trim piece to cover any gaps around the fan opening ❷.

1 Apply a bead of caulk under the flanges before you set the fan box in place.

2 Install the trim pieces to cover the gaps between the fan and the ceiling cutout.

TRADE SECRET

If there's an existing ceiling light in the bathroom, a fan-and-light combination unit may simplify your remodel. If you want the fan and light to operate at the same time, you can reuse the two-wire cable that's presently controlling the light switch. But if you want to operate the fan and light separately—so that the fan continues running after the light is switched off—you'll have to install a length of 14/3 or 12/3 cable, as shown on p. 155.

WIRING A BATHROOM FAN

1 Staple cable no more than 1 ft. from the fan box.

2 Strip the sheathing off the wires.

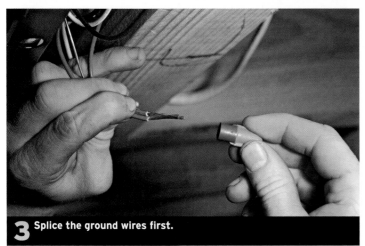
3 Splice the ground wires first.

4 Fold the spliced wires into the junction box.

Before making any connections in the fan's junction box, make sure the power is off. Follow the wiring diagrams provided by the fan manufacturer. In general, it's easier if the incoming power runs through a switch box first; that you don't have to try to splice the switch legs in the fan junction box—junction boxes inside fan housings tend to be cramped. Bathroom fans should also have ground-fault circuit interrupter (GFCI) protection if installed in a shower whose ceilings are 8 ft. above the finish floor or lower.

If the duct space is tight, it's usually best to wire the fan box before installing the duct. When running cable to the fan box, allow a generous loop of cable, just in case. As with light fixtures and receptacles, staple the cable within 1 ft. of the fan box ❶. Insert a cable connector into a junction box knockout, feed the cable through it, and strip the cable sheathing ❷.

Using wire connectors, first splice the incoming ground wire to the fixture ground ❸ (If the fixture lead wires are stranded, extend them slightly beyond the solid wire so that the wire connector will engage them first). Splice the neutral wires and then the hot wires. If the light and fan are wired separately, there will be two sets of hot wires.

Tug each wire group gently to be sure the splices are secure. When all groups are spliced, carefully fold the wires into the fan junction box and cover the box ❹.

TRADE SECRET

In tight spaces like a junction box, almost all the pros use a utility knife to remove sheathing from the wires.

RANGE HOODS

Choosing a Range Hood

All cooktops and stoves should be vented by a range hood. In addition to sucking up the smoke of a charred steak, range hoods exhaust airborne grease that might otherwise migrate to a cool corner and feed mold or adhere to woodwork and discolor its finish. Range hoods come in many configurations but basically there's a *hood* to collect smells and smoke, a *fan* to expel them, *ducts* to carry exhausts out, and *shrouds* and other trim pieces plus, of course, wires, switches, lights, and whatnot. Range hoods vary from low-powered and inexpensive (less than $50) to custom-designed units (which cost thousands).

Range hoods are most often *wall-mounted* directly over a range. Alternatively, there are *downdraft* and *side-draft* vents that pop up from a counter area to suck away fumes. Over island and peninsula ranges, you can install *chimney-type* vents. In general, install the type of vent that will carry exhaust gases outdoors with the shortest and straightest duct run possible. Because heated air rises, wall-mounted and chimney types are inherently efficient, whereas downdraft and side-draft vents pull heated gases in directions they wouldn't go naturally and can even pull burner flames sideways.

Ideally, a range hood should be slightly wider than the range, say, 3 in. wider on each end, and mounted 30 in. above the range, but follow the hood maker's suggested mounting height. More powerful hoods can be installed higher. Finally, buy a unit with a good-quality filter that can withstand regular washing with soap and water. Most filters are aluminum mesh, better ones are stainless steel; many can be popped into the dishwasher, which spares homeowners a very greasy and unpleasant task. In general, be skeptical of range hoods that recirculate air through a series of filters rather than venting it outside.

Range hoods can be handsome additions to kitchen decor. This one is a European-style wall-mounted hood from Kobe®.

The height of the fan and the fact that it is not against a wall mean that the hood needs to draw more air to be effective. This stylish design is set higher than a typical range hood to provide sight lines across the room.

SIZING THE HOOD

A 100-cfm wall-mounted hood should be adequate to vent the average four-burner, 30-in.-wide range. But if that same range is located on a kitchen island, its range vent should draw 125 cfm to 150 cfm. More is not better when sizing range hoods. For one thing, larger hoods are noisier. Midsize range hoods average 3 sones to 3.5 sones (a measure of noise), which is too noisy to have a conversation nearby; monster hoods can reach 8 sones. (In comparison, refrigerators register 1 sone.) Oversize hoods can also expel so much air that they create *back-drafting,* in which negative in-house air pressure draws furnace or fireplace exhaust gases back down the chimney.

MOUNTING & DUCTING A RANGE HOOD

1 Slide the duct into the thimble.

2 Level, center, and mount the hood.

3 Bolt at least one side of the hood to a stud.

4 Lower the duct over the hood takeoff and seal the joint with metal tape.

> **(i) TRADE SECRET**
> A well-sized hood should extend about 3 in. beyond the range on both ends.

Range hoods are typically screwed to the underside of a cabinet or mounted directly to a wall. A hood should be mounted about 30 in. above the range, and the ducting that vents it should exit the house as directly as possible. As you plan your duct route, drill pilot holes to locate studs in the wall behind the stove or through ceiling joists above. The duct run should exit through the siding or the roofing. To locate the hole for the duct work, level the hood and center it over the range.

Cut a hole in the wall or ceiling and insert a thimble—a specialized fitting that creates an opening through a wall or ceiling into which the duct fits. If there already is a thimble in place, insert the duct from the range hood up into the thimble 3 in. to 4 in. above its final position **1**. Friction should hold the duct in place.

Level the hood, center it over the range, and bolt it to the wall **2**. Mount the hood in at least two points: If the studs behind the

hood are 16 in. o.c., you may have to bolt one side of the hood to a stud **3** and secure the other side of the hood with a toggle bolt. After the hood is mounted, slide the duct down over the takeoff atop the hood **4**. Tape all of the joints with self-adhering metal tape.

WIRING THE RANGE HOOD

Many range hoods have discrete electrical junction boxes with knockouts, into which you insert cable clamps and cables. In the example shown here, the shroud that encloses the duct doubles as a junction box. The lead wires from the fan and light emerge through a bushing on the top of the hood and attach to wires in a Romex supply cable, which emerges from the wall cavity.

Strip and splice like wire pairs using wire connectors. Splice the ground wires first—be sure there's a grounding pigtail to the hood itself—then neutral wires, and then hot wires ❶.

To minimize weight and avoid marring shiny trim pieces, many pros install the fan blades, filters, trim, and other trappings after the hood shell is mounted. Most of the parts simply snap into place—just follow the instructions provided ❷. Then, if your hood has a bottom casing, screw it into place ❸. Finally, fit the shroud over the ducting and the wire connection—they're usually held in place with one pair of screws at the top and another pair at "the bottom ❹.

WARNING
Most hoods aren't heavy but they're unwieldy, so get help installing one.

1 Splice the grounds, then neutrals, and then hot wires.

2 Snap on the fans, filters, and trim pieces.

3 Attach the bottom casing onto the hood.

4 Use screws to secure the shroud in place.

CEILING FANS

Choosing a Ceiling Fan

These days, you can find fan styles that match almost any architectural style; the Internet is a good place to start looking. As you might expect, quality correlates closely to cost. Good-quality fans will be balanced, so they don't wobble, and will run quieter, thanks to features like sealed steel bearings.

Most fan companies offer three mounting profiles: *standard* (which employs extension downrods to situate fan blades at an optimal height), *flush* (for low ceilings), and *angled* (for sloping ceilings). To operate safely, fan blades must be located at least 7 ft. above the floor; optimally, fan blades should be 8 ft. to 10 ft. above the floor. To attain that height from ceilings higher than 9 ft., fan makers offer extension downrods of varying length. Most downrods are $1/2$ in. in diameter; better quality units use rods that are $3/4$ in. in diameter. For safety reasons, fan blades must be at least 18 in. away from the nearest wall or sloping ceiling.

When spinning, ceiling fan blade spans range from 32 in. to 60 in. There are, of course, complicated formulas for determining fan size, and ideally you should center any fan over the area in which people are most likely to congregate in a room. The chart on p. 155 will help you determine the right size.

In general, more fan blades won't move more air. More important is the blade angle. The steeper the pitch of the blade, the more air it will move. Less expensive fan blades will have a 10-degree pitch and spin faster; better-quality fans will have a 12-degree to 14-degree pitch.

Another important consideration is the switch that controls the fan. Most fans have a multiple-speed pull-chain switch built into the bottom of the housing. If the fan is too high for you to reach the chain comfortably or you just don't like the look of a pull chain, install a wall switch to control the fan. Among the more popular fan controls are three-speed wall switches and remote wireless controls. Avoid using dimmer switches (rheostats) to control the fan, however, as rheostats frequently cause fan motors to hum or, in some cases, burn out. Follow manufacturer's recommendations for appropriate switches.

This low-profile Nouvelle® fan comes with a bronze patina and walnut blades.

Remote controls, like this Adapt-Touch™, allow you to forgo long pull chains.

CEILING FAN LIGHTS

Many homes are designed with a single light outlet in the center of each room. If homeowners want both a fan and a ceiling light in a given room, installing a ceiling fan light is a straightforward and cost-effective solution. Thus many fan makers offer light kits, or fans with integral lights. Until recently, this all-in-one solution resembled a man wearing two hats because the light looked glued onto the fan. Fortunately, some fan makers are now designing combo units that look more integrated—search for them, they're out there. If you still don't find a combo fan and light that fits your taste, add a separate lighting outlet.

The integrated LED light fixture of this Scandia® fan can be modulated by a dimmer switch.

INSTALLING THE CEILING FAN BOX

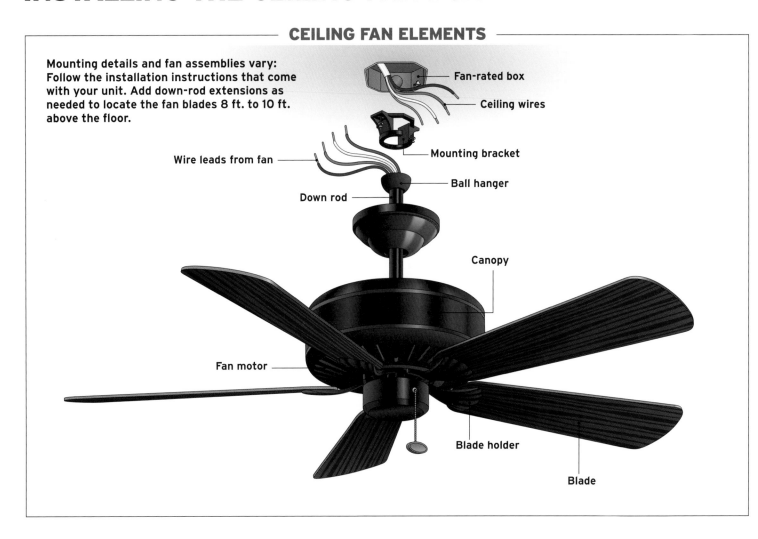

CEILING FAN ELEMENTS

Mounting details and fan assemblies vary: Follow the installation instructions that come with your unit. Add down-rod extensions as needed to locate the fan blades 8 ft. to 10 ft. above the floor.

Fan-rated box

Ceiling wires

Wire leads from fan

Mounting bracket

Ball hanger

Down rod

Canopy

Fan motor

Blade holder

Blade

WARNING

Although there are plastic ceiling boxes rated for fans, many installers won't touch them. Spooked by the possibility of stripped screw holes and crashing fans, they insist on using a metal fan box.

All fans must be securely mounted to framing, but this is especially true of ceiling fans, which are weighty and subject to vibration and wobbling. All fans must be installed in a ceiling box rated for fan use—that is, capable of supporting 50 lb. or more. Check your fan's installation specs. If your fan is particularly heavy or complex, it may require a box with a higher weight rating or additional bracing.

If ceiling joists are exposed, you have several options. You can install a piece of 2x4 or 2x6 blocking to the ceiling joists (or rafters), then mount a 2-in.-deep octagonal metal box to the blocking. Use at least three 3-in.-long wood screws on each end of the blocking. Adjust the blocking's depth so the box will be flush to the finish ceiling.

If you must hang a fan from a ceiling joist's edge, screw a 1/2-in.-deep metal pancake box directly to the joist. Before doing so, however, be sure there is enough room beneath the fan's canopy to hide electrical connections—because there's no room for them in the pancake box! If not, move the fan or modify your plans.

An alternative is to mount an adjustable hanger to the framing. Bar hangers are strong, easier than nailing up blocking, and easily positioned by sliding a box along its support bar. *Some types of hangers can also be used as remodel bars.*

➡ **For more on bar hangers, see p. 29.**

If joists are exposed, **install the blocking and mount a fan box.**

1 Insert a remodel bar through the ceiling cutout.

2 Expand the remodel bar by hand.

3 Attach the fan box to the remodel bar.

WARNING

If there's an existing fan box, turn off the power to it and test before proceeding. If the box is plastic, check the screw holes that the fan mounting bracket attaches to. If the screw holes are at all stripped, of the box is cracked or deformed, or if you have any doubts that it can support the new fan, replace the box.

If joists are not exposed, install a remodel bar (AKA a *braced box*) if there's a finish ceiling. Locate the fan, cut a 4-in.-diameter hole in the ceiling, insert the bar into the hole **❶**, and maneuver it till its feet stand on top of the drywall. Then hand-turn the bar to expand it **❷**. When the bar touches a joist on both ends, use a wrench to drive the bar points into the joists. Finally, bolt the ceiling box to the remodel bar using the hardware provided **❸**.

MOUNTING THE CEILING FAN

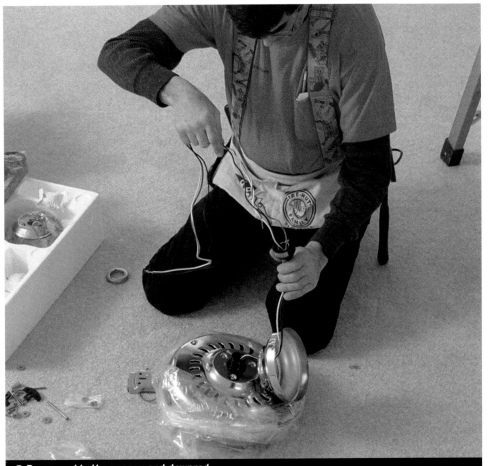

1 Preassemble the canopy and downrod.

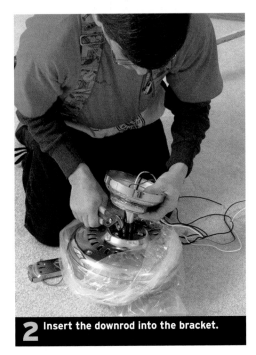

2 Insert the downrod into the bracket.

3 Bolt the blade holders to the fan motor.

Fan assemblies differ greatly, so be sure to follow the installation instructions that come with your fan. After testing to be sure the power is off, feed the ceiling wires through the opening in the mounting bracket and screw the bracket to the fan box. *Note:* For some assemblies, the mounting bracket is simply a flat crossbar that spans the fan box; for other types, the fan's weight is supported by a bracket that is somewhat like a slotted dome, into which a ball hanger fits.

Preassemble the parts necessary for hanging the fan motor and housing. Typically, you feed the fan's wire leads up through a canopy and a hollow downrod. The downrod has a ball head that slips into, and is supported by, a slotted mounting bracket screwed to a ceiling outlet box **❶**.

Carefully lift the fan motor toward the mounting bracket in the ceiling. Being careful not to pinch the wires from the ceiling or those from the fan motor, insert the ball head of the downrod into the opening of the mounting bracket **❷**. On some models, you may need to rotate the fan body until a slot in the ball head aligns with a pin in the bracket. If there is a locking bolt or hanging pin that secures the assembly, be sure to attach it or the unit may fall.

Now wire the unit. Because all metal boxes must be grounded, attach a grounding pigtail to the bracket. Use wire connectors to join like wire pairs, starting with grounding wires, then the neutral wires, and then hot wires. If there's a fan light, there will be two hot wire leads: The black lead typically connects to the fan and the blue or red lead controls the light. Tuck the spliced wires behind the canopy, slide the canopy up the downrod until the canopy is flush to the ceiling, then tighten the setscrew that holds it in place.

Attach the fan blades last. Typically, four machine screws attach a blade to each blade holder; each screw hole is lined with a rubber grommet that cushions the screw-and-nut assembly and forestalls vibration. Each blade holder then bolts to the underside of the fan motor ❸. Because fan blades are prebalanced at the factory, never use blades from other fans. Once you've tightened and checked all bolts and screws, turn the fan on its slowest speed to check for wobbling.

TRADE SECRET

The most common cause of wobbling is a blade holder that's become misaligned. To check for that, turn off the fan and use a yardstick to see if all blade holders are the same height from the ceiling. Most blade holders can be removed and bent slightly back into alignment; the store that sold you the fan may also have a technician. Fan makers also have websites that describe how to balance and troubleshoot their fans.

WIRING A FAN-LIGHT COMBINATION

The fan types discussed in this section are frequently configured with both fans and lights. Running a three-wire cable (12/3 with ground or 14/3 with ground) enables you to operate the fan and light separately. When a ceiling fan's junction box is closer to the power source than to the switch box, as shown in the drawing below right, the switch is called a "back-fed switch." When wiring a back-fed switch leg, tape the white wire black to indicate that it's being used as a hot wire.

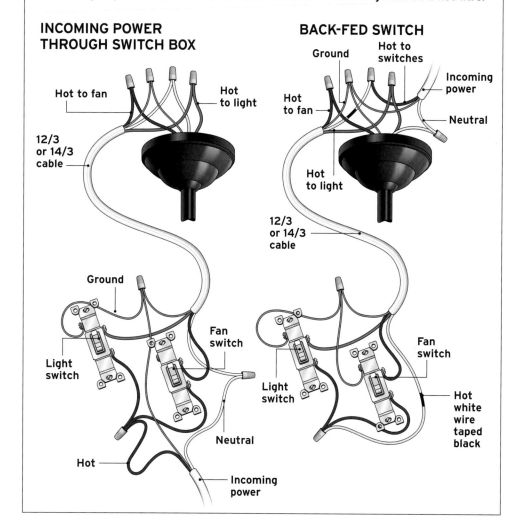

INCOMING POWER THROUGH SWITCH BOX

Hot to fan
Hot to light
12/3 or 14/3 cable
Ground
Fan switch
Light switch
Neutral
Hot
Incoming power

BACK-FED SWITCH

Ground
Hot to switches
Hot to fan
Incoming power
Neutral
Hot to light
12/3 or 14/3 cable
Fan switch
Light switch
Hot white wire taped black
Incoming power

FAN SIZES	
Blade Span (in.)	Room Size (sq. ft.), max*
32 or 36	100
42 or 44	144
48	225
50 or 52	400
54 to 60	> 400

* Output may also be expressed in cfm (cubic feet per minute). For example, a good-quality fan with a 36-in. span will move 2500 cfm to 4000 cfm.

PLANNING NEW WORK

BEING DETAIL ORIENTED IS AS important to planning as it is to installation. When you plan a wiring project, be methodical: Assess the existing system, calculate electrical loads, check local codes, and draw a wiring floor plan.

If you are only replacing existing devices—changing a light fixture, replacing a faulty switch, or upgrading a receptacle, for example—you seldom need a permit from the local building department. However, if you extend or add any circuit, you must pull (or file) a permit.

Most local electrical codes are based on the NEC. When it's necessary to pull a permit, local code authorities will want to approve your plans and later inspect the wiring to be sure it's correct. Don't short-circuit this process: Codes and inspections protect you and your home.

Whatever the scope of your project, if you work on existing circuits, first turn off the power and test to be sure it's off, as shown throughout this book.

BEFORE YOU BEGIN

Inspecting the fuse box or breaker panel, p. 158

Assessing wiring condition, p. 160

Is the system adequately sized?, p. 161

ACCORDING TO CODE

Common code requirements, p. 163

General-use circuit requirements, p. 164

WORKING DRAWINGS

Developing a floor plan, p. 165

Electrical notation, p. 168

Receptacles, p. 170

Switches & lights, p. 172

Receptacles, switches & lights, p. 174

INSPECTING THE FUSE BOX OR BREAKER PANEL

A This 30-amp main has seen better days. Even if it were safe, it would be dramatically undersize for today's electrical needs.

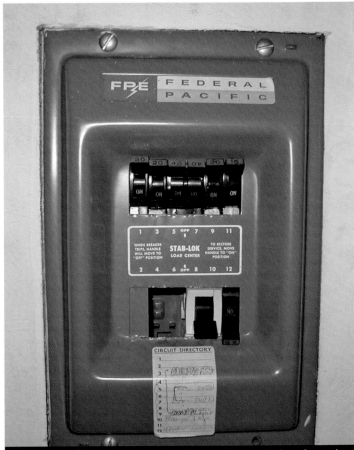

B Replace this panel: Federal Pacific panels are notorious for arcing and panel fires and hot bus bars are exposed.

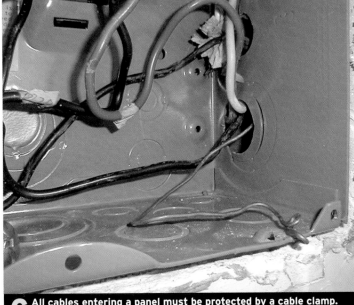

C All cables entering a panel must be protected by a cable clamp. Sharp panel edges could slice wire insulation.

D Here, service cable enters a panel without a cable connector, and the box edge has nearly sliced through the sheathing.

By looking at the outside of the service panel and wiring that's exposed in the basement and attic, you can get a basic overview of the system's condition. If the wiring is in decent shape, you can probably continue using it and safely add an outlet or two. However, if the system seems unsafe or inadequate, hire a licensed electrician to open the panel and do a more thorough examination. Here's what to look for.

Start your investigation at the fuse box or breaker panel. You can learn a lot about the condition of the system by examining the outside of the service box. Examining the inside of a panel or fuse box is best left to a licensed electrician, however.

➤ **For more on inspecting the panel, see p. 14.**

Rust and corrosion on the outside of a service box or on the armored cable or conduit feeding it, can indicate corroded connections inside **Ⓐ**. Such connections can lead to arcing and house fires, so have a licensed electrician replace the fuse box or panel. Likewise, if you see scorch marks on breakers or a panel, have a pro examine it.

Melted wire insulation is a sign either of an overheated circuit—usually caused by too many appliances in use at the same time—or of a poor wire connection in which arcing has occurred. In the first case, a homeowner typically installs an oversize fuse or breaker to keep an overloaded circuit from blowing so often; but this "remedy" exceeds the current carrying capacity of the wire. The wire overheats and melts its insulation, which can lead to arcing, house fires, or—if someone touches that bare copper wire—electrocution.

A fuse box full of 20-amp or 30-amp fuses may not melt wires where you can see them, but it may have damaged wire insulation somewhere you can't. This is called overfusing. Have an electrician inspect the electrical system. Installing type-S fuse socket inserts can prevent overfusing.

"Pennying" a fuse is another unsafe way to deal with an overloaded circuit that keeps blowing fuses. In this case, someone unscrews a fuse, inserts a penny or a blank metal slug into the bottom of the socket—a dangerous act in itself—and then reinstalls the fuse. The penny allows current to bypass the fuse and the protection it offers. Here, again, have an electrician examine the circuits for damage to the wire insulation.

Panel covers that don't fit, have gaps, or are missing are unsafe. Likewise, any installation in which panels,

covers, and breakers are from different manufacturers is a Code violation; the NEC requires that panel components be listed (UL listed, for example) and come from a single manufacturer. So if you see covers that have been cut to fit a breaker, cover knockouts that are missing, bus bars that are visible when the panel cover is on, or mismatched components, -hire a licensed electrician to correct those problems. Some older brands of breaker panels, such as Federal Pacific® and Zinsco®, have spotty reputations and should also be replaced **Ⓑ**.

Missing cable connectors or unfilled knockouts enable mice and vermin to enter the panel and nest in it, which can be a fire hazard. Missing connectors also allow cables to be yanked, stressing electrical connections inside the panel **Ⓒ**. A missing cable clamp may also allow the sharp edge of the panel to slice through thermoplastic cable sheathing, which could energize the panel and electrocute anyone who touched it **Ⓓ**.

An ungrounded service panel, a major hazard, is explained in greater length in Chapter 1. A properly grounded panel will have a large grounding wire running from the panel to a grounding electrode (rod). For the entire electrical system to be grounded, there must be continuous ground wires or other effective grounding path running from each device or fixture to the service panel and by extension to the grounding electrode. Cold water and gas pipes must also be connected (bonded).

➤ **For more on grounding, see p. 12.**

ASSESSING WIRING CONDITION

Cables may be visible as they near the service panel and as they run through attics and basements. If there are covered junction boxes, carefully remove the covers and examine the wire splices inside—without touching them. You can also turn off power and pull a few receptacles out to better examine the wires.

Deteriorated sheathing is a potential shock hazard, so note brittle fiber insulation and bare wire, but avoid touching it. Sheathing that's been chewed on by mice, rats, or squirrels should be replaced.

NM cable must be stapled within 12 in. of boxes and every 4½ ft. Sagging wire is hazardous because it can get inadvertently strained, jeopardizing electrical connections. Likewise, all boxes must be securely mounted. All NM cable entering metal boxes must be gripped by cable clamps; whereas plastic boxes do not require the strain relief of cable clamps.

Aluminum circuit wiring is a fire hazard unless it is correctly terminated with a COPALUM connector or CO/ALR-rated outlets and switches. If it is incorrectly terminated in a copper device, the two metals will expand and contract at different rates each time the circuit is under load. This can lead to loose connections, arcing, overheating, and house fires.

Wire splices must be housed within a covered junction box, outlet box, or the like. Wires that are spliced outside a box or inside an uncovered box can be a fire hazard because of the dangers of arcing (sparks leaping gaps between wires). Loose connections not contained in a cover box can easily ignite combustibles nearby because arcs approach 2,000°F.

Knob-and-tube wiring, although outdated, is inherently safe unless individual wire insulation is deteriorated or splices are incorrectly made. Typically, splices that were part of the original installation will not be housed, but must be wrapped with electrical (friction) tape and supported by porcelain knobs on both sides of each splice. Non-original splices must be housed in covered boxes. Have knob-and-tube wiring assessed or modified by an electrician familiar with it; it's quirky stuff. The NEC does not allow knob-and-tube wiring to be buried in insulation, though some local inspectors are okay with the practice.

Unstapled cable can be inadvertently yanked and stress electrical connections. If you see work this sloppy, suspect substandard wiring throughout the house.

Ungrounded outlets

You can continue using two-slot, ungrounded receptacles on general-use circuits as long as the wires are in good condition and the connections are correctly made. Most new lamp and appliance plugs are double-insulated so there's almost no risk of getting a shock from the plug. However, if you want to use appliances or tools with three-pronged plugs or a surge protector at the location, you must upgrade to a grounded circuit. Putting a two-pronged adapter on a three-pronged plug is unsafe!

A PRO'S TAKE ON REWIRING

If an older home's wiring is in decent shape, it's probably okay to continue using it, even though it may not meet electrical code for a new installation. If you are planning to gut the house completely, it might make sense to rip out all the old wiring and completely rewire the house. But if you're remodeling only part of the house, leave most of the old wiring in place and spend your money rewiring the kitchen, baths, and laundry circuits. That will give you more bang for your buck.

However, you should replace old wiring that's unsafe. If you observe any of these conditions, the wiring should be replaced:

- Circuits that have been extended improperly, as evidenced by loose connections, unprotected splices, or arcing.

- Knob-and-tube wiring whose insulation has been damaged. Also, if knob-and-tube wiring in the attic has been covered with loose-fill insulation or insulation batts, that is a serious Code violation that could lead to overheating and fire danger—that wiring should be replaced.

- Circuits wired with unsheathed wires rather than with sheathed cable or conduit.

IS THE SYSTEM ADEQUATELY SIZED?

If receptacles in your house teem with multiplugs and extension cords, you may need to add more outlets. But there are also more subtle clues: If you blow fuses or trip breakers regularly, or if the lights brown out when you plug in a toaster or an electric hair dryer, you've got overloaded circuits and may need to add new circuits to relieve the load on existing circuits. This section will help you figure out whether your system has the capacity to add new circuits or add outlets to existing circuits.

Electrical service, revisited

Let's start with a recap of the electrical service running from the utility pole to your house. If there are only two large wires running from the utility pole to the house, they deliver only 120v service. A house with two-wire service probably has a 30-amp or 60-amp main fuse, which is inadequate for modern usage.

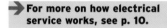
For more on how electrical service works, see p. 10.

These days, three-wire service feeding a 100-amp service panel is considered minimal, and many electricians install 150-amp or 200-amp panels if the home-owners plan to enlarge the house at some point or acquire a lot of heavy energy-using appliances such as electric ranges and hot tubs.

Offhand options

The only sure way to know if you've got enough capacity to add an outlet or a circuit is to calculate electrical loads.

For more on calculated electrical loads, see p. 162.

But for the benefit of those who wish that an electrician would just offer an offhand opinion of what works most of the time, the "rules of thumb" below continue on the next page.

Fuse box service If you've got a fuse box with a 30-amp or 60-amp main fuse, the best advice we can give is: Get rid of it. Don't add outlets or circuits until you replace the outside service box with a breaker panel. A fused main should be replaced because it has a limited load capacity, it is frequently abused by people trying to bypass its protection, and it is hated by

>> >> >>

Three-wire service, made up of two large, insulated 120v wires and a bare ground/neutral wire, supplies enough power for modern needs.

VOLTS, AMPS, AND WATTS

At a power plant or a substation, electricity is multiplied (charged) and given pressure (*voltage*); in that form, electricity is potential energy, just like a charged battery. When electricity is put to work at an outlet, electrons flow through the wires, and power is delivered as heat or light. The *ampere* (amp) is a unit in which this current flow is measured. The amount of energy consumed at a given point—say, at a toaster or a light bulb—is measured in *watts*. Volts, amperes, and watts are thus interrelated:

- *Voltage:* the potential to do work (electrical pressure).
- *Amperes:* the rate of electrical flow.
- *Watts:* the rate at which energy is consumed.

Or, expressed as mathematical formulas:

- Watts = Voltage x Amperes
- Amperes = Watts ÷ Voltage

To reiterate briefly, electricity, impelled by voltage, flows from the power source. Along the way (at outlets), it encounters resistance and does work. It then returns to the power source, its voltage reduced or spent.

To determine the load-bearing capacity of a circuit you want to extend, identify the circuit breaker controlling the circuit and note the rating of the breaker. If it's a general-purpose circuit, the breaker will probably be 15 amp or 20 amp. A circuit controlled by a 15-amp breaker has a capacity of 1,800w (15 amp x 120v); one controlled by a 20-amp breaker, has 2,400w.

The total wattage of all energy users on the newly extended circuit must not exceed these capacities; otherwise, you risk overheating wires. To avoid overloading, load is calculated at 80 percent of capacity. For example, 80 percent of 1,800w is 1,440w for a 15-amp circuit; 80 percent of 2,400w equals 1,920w for a 20-amp circuit. As a rule of thumb, 10 outlets is the maximum for a general-purpose or lighting circuit.

CIRCUIT CAPACITIES		
Amperes x Volts*	Total Capacity (watts)	Safe Capacity† (watts)
15 x 120	1,800	1,440
20 x 120	2,400	1,920
25 x 120	3,000	2,400
30 x 120	3,600	2,990

*Safe capacity = 80 percent of total capacity. †Amperes multiplied by volts equals watts.

IS THE SYSTEM ADEQUATELY SIZED? (CONTINUED)

insurance companies. Upgrade the service, and your insurance premiums may decrease. In many setups there will also be a fuse box inside, which distributes power to house circuits. You can leave the inside fuse box in place.

Adding outlets If you've got a breaker panel, you can almost always add an outlet or two—a receptacle or more lights, for instance. If, for example, you have three-wire service and a 100-amp main, there's usually a lot of excess capacity.

Adding a circuit for general use If there's space in the panel to remove a knock-out and add a breaker, go ahead. This may be necessary if you're adding a home entertainment center or a computer. Computers and home entertainment centers, however, aren't huge energy users.

Adding a kitchen or bath circuit
Here, the answer is not such a slam dunk. First, see if there's space in the panel to add a breaker. If you're adding a bath fan or some new light fixtures, no problem. If you're adding a 20-amp, small-appliance circuit to reduce the load on an existing circuit, you're probably okay.

Remodeling a kitchen Kitchens are complicated and often full of big energy users. Use the chart at right to help you add up the loads. If there aren't many open spaces for breakers, you may need to upgrade to a larger panel.

Adding dedicated circuits If you need to add dedicated circuits for heavy-use items such as an electric range (50 amps) or a hot tub (60 amps), get out the calculator and do the math.

LOAD CALCULATION SINGLE FAMILY DWELLING

1. GENERAL LIGHTING LOAD

Type of Load	NEC Reference	Calculation	Total VA
Lighting Load	Table 220-3 (b)	sq.ft X 3 VA	0.00 VA
Small appliance Load	Section 220-16 (a)	circuits X 1500 VA	0.00 VA
Laundry Load	Section 220-16 (b)	circuits X 1500 VA	0.00 VA
Total General Lighting			VA

2. DEMAND FACTOR FOR GENERAL LIGHTING LOAD

Type of Load	Calculation	Demand Factor	Total VA
General Lighting (1)	First 3000 VA X	100%	3000 VA
General Lighting (1)	Remaining VA X	100%	VA
	Total VA (demand)	=	VA

Total Lighting, Small Appliances & Laundry (A)

3. LARGE APPLIANCES

Type of Load	Nameplate Rating	Demand Factor	Total VA
Electric Range	Not Over 12KVA	Use 8KVA	VA
Clothes Dryer	VA	100%	VA
Water Heater	VA	100%	VA
Other Fixed Appliances	VA	100%	VA
Total Load for Large appliances (B)			——VA

	Total VA (A+B)	———	0.00 VA
Minimum Service Size	Total VA /240V	———	0.00 VA

USING THIS CHART

1. Square ft. for general lighting load is for the entire dwelling including habitable baasements or attics.
2. NEC requires a minimum of 2 small appliance loads, but it is important to add small fixed kitchen appliances (microwaves, disposals, etc.).
3. Minimum of 1 laundry load is required for a single family dwelling.
4. The demand factor calculation is designed to take actal use into account (e.g. it is unlikely all lights and small appliances will be running at one time).
5. All large load appliances (high wattage) are added at 100%.
6. The final load calculation in the minimum. Often increasing capacity has little cost impact and is a good practice.

COMMON CODE REQUIREMENTS

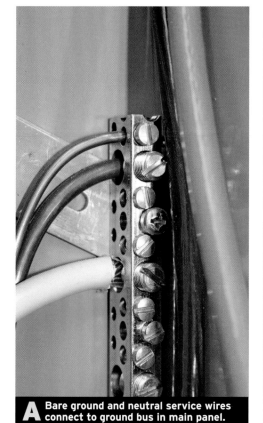

A Bare ground and neutral service wires connect to ground bus in main panel.

B Large copper ground wire is clamped to grounding electrode (rod).

C The main water pipe should be properly grounded.

D All metal boxes and appliance housings require grounding.

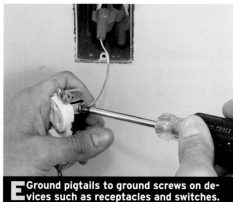

E Ground pigtails to ground screws on devices such as receptacles and switches.

Electricians follow the NEC, which was compiled to promote safe practices and prevent house fires. Consider these requirements before you start drawing plans, but be sure to consult local electrical code, too—it is the final authority in your area.

The requirements given here apply to all circuits in the house, whether general lighting or heavy-use appliance circuits. Local codes rarely require you to change existing circuits—as long as they are safe—but new electrical work should reflect current electrical code.

Circuit wiring

Wire gauge must be large enough to carry the circuit load and be protected by a comparably sized breaker or fuse at the panel. General-use and lighting circuits are typically 14AWG wire, protected by a 15-amp breakers; kitchen, bath, and workshop circuits usually have 12AWG wire, protected by 20-amp breakers.

➔ See the facing page for more on calculating circuit loads.

Acceptable cable

Most circuits are wired with NM Romex cable because they are protected behind finish surfaces. When circuit wiring is to be left unprotected and exposed, it must be armored cable or conduit.

➔ For more on cable and conduit, see p. 204.

Grounding

All receptacles, appliances, and electrical equipment must be separately connected (bonded) to the service panel. NM cable contains a separate ground wire, whereas armored cable sheathing and metal conduit provide the path to ground **A**-**E**.

Boxes

All electrical connections must take place in covered boxes. Based on local code requirements, boxes may be plastic or metal. If metal, the box must also be separately connected (bonded). If NM cable is used, the ground wire must be connected to the box with either a ground screw or a ground clip. If AC cable or metal conduit is used, it must be properly attached to the box to ensure effective bonding. If the box is plastic, it does not need to be (and cannot be) grounded; run a ground wire to the device or fixture only.

GENERAL-USE CIRCUIT REQUIREMENTS

General-use circuits are intended primarily for lighting; but small users, such as televisions, fans, and vacuums, are allowed—as long as the loads they draw don't exceed the capacity of the circuit.

Lighting and small users Though 14AWG wire is sufficient for lighting and switch runs, electricians often run 12AWG wire on general-use circuits to accommodate future uses. Calculate lighting loads at 3w per square foot, or roughly one 15-amp circuit for every 500 sq. ft. of floor space. When laying out the lighting circuits, do not put all the lights on a floor on one circuit. Otherwise, should a breaker trip, the entire floor would be without power.

Receptacles There must be a receptacle within 6 ft. of each doorway, and receptacles should be spaced at least every 12 ft. along a wall. (This is also stated as, "No space on a wall should be more than 6 ft. from a receptacle.") Any wall at least 2 ft. wide must have a receptacle; and a receptacle is required in hallway walls 10 ft. or longer.

Outlets The NEC does not specify a maximum number of outlets on a residential lighting or appliance circuit, though local jurisdictions may. Figure roughly 9 outlets per 15-amp circuit and 10 outlets per 20-amp circuit.

Light switches There must be at least one wall switch that controls lighting in habitable rooms, the garage, and storage areas (including attics and basements). There must be a switch near each outdoor entrance. Three-way switches are required at each end of corridors and at the top and bottom of stairs with six steps or more. When possible, put switches near the lights they control. It should be noted that the light switch can control a receptacle (considered "lighting") in habitable rooms.

AFCI protection All new 15-amp and 20-amp circuits in bedrooms must have arc-fault circuit interrupter (AFCI) breaker protection to guard against house fires.

KITCHEN LIGHTING BASICS

Kitchen lighting should be designed to utilize natural light during the day and be a balance of general light and task lighting at night. Do not be afraid of energy efficient lighting such as fluorescent. Today's energy efficient lighting is instant, dimmable, and available in colors that match incandescent light. Kitchen lighting is often highly regulated for energy efficiency. Check with local building officials before you begin your design.

GENERAL LIGHTING

General lighting is meant to illuminate the space generally and can come from recessed cans, surface mounted fixtures, track lighting, or cove uplighting. Consider cabinetry and appliances when layout out new light fixtures. A general rule is 2w incandescent or 1w fluorescent per square foot of kitchen area, but even illumination is the goal.

TASK LIGHTING

Task lighting is meant to provide a higher level of illumination at work areas (sinks, counter tops, and islands) and can be achieved with recessed cans, pendants, or undercabinet fixtures. If cabinets are over countertops, undercabinet fixtures (T5 fluorescent or halogen strips) are by far the best choice and should be spaced for even illumination of the counter surface. For islands and sinks choose a recessed can with a slightly higher wattage and narrower lamp beam spread, or install pendants with similar attributes.

An AFCI breaker can detect **the minute current fluctuations associated with arcing. It then shuts off power to protect you from house fires.**

DEVELOPING A FLOOR PLAN

D rawing a set of project plans can help you anticipate problems; find optimal routes for running cable; minimize mess and disruption; and in general, maximize your time and money. A carefully drawn set of plans is also an important part of the code compliance and inspection process.

If you're replacing only a receptacle, switch, or light fixture, you usually don't need to involve the local building department. But if you run cable to extend a circuit, add a new circuit, or plan extensive upgrades, visit the building department to learn local code requirements and take out a permit.

As noted throughout this book, the NEC is the foundation of national and local electrical codes for homes and businesses. Local code authorities have the final say, however; so your wiring plans should be approved by a local building inspector before you start the project.

Phone first

Call the building department and ask if local codes allow homeowners to do electrical work or if it must be done by a licensed electrician. You may be required to take a test to prove basic competency. This is also a good time to ask if the municipality has pamphlets that give an overview of local electrical code requirements. >> >> >>

REQUIREMENTS
ROOM BY ROOM

Kitchen and bath appliances are heavy power users, so their circuits must be sized accordingly.

GFCI PROTECTION
The NEC requires GFCI protection for all bathroom receptacles; all receptacles serving kitchen counters; all outdoor receptacles; accessible basement or garage receptacles; and receptacles near pools, hot tubs, and the like. (Check the current NEC for a complete listing.)

BATHROOM CIRCUITS:
Bathroom receptacles must be supplied by a 20-amp GFCI protected circuit. The NEC allows the 20-amp circuit to supply the receptacles of more than one bathroom or to supply the receptacles, lights, and fans (excluding heating fans) in one bathroom. New or remodeled bathrooms must have a vent fan.

SMALL-APPLIANCE CIRCUITS
There must be at least two 20-amp small-appliance circuits in the kitchen. No point along a kitchen countertop should be more than 2 ft. from an outlet—in other words, space countertop receptacles at least every 4 ft. Every counter at least 12 in. wide must have a receptacle.

KITCHEN LIGHTING:
Adequate lighting is particularly important in kitchens so people can work safely and efficiently. Layout a good balance of general and task lighting. Be aware that many jurisdictions have energy efficiency requirements for lighting in kitchens so check with your local building authority first.

BATHROOM LIGHTING:
It is important to illuminate the face evenly in mirrors. Common practice is to place good quality light sources either above the vanity mirror or on either side of it. Be careful when using recessed cans over the vanity for they can leave shadows across the face. Many jurisdictions also have energy efficiency requirements for lighting in bathrooms including lighting and occupancy sensors.

DEDICATED CIRCUITS
All critical-use stationary appliances must have their own dedicated (separate) circuits. Critical-use appliances include the water pump, freezer, refrigerator, oven, cooktop, microwave, furnace and/or whole-house air-conditioning unit, garbage disposal, window air-conditioners, and water heater. A bathroom heater requires a dedicated circuit, whether it is separate unit or part of a light/fan. Laundry room receptacles must be on a dedicated circuit, as should an electric clothes dryer.

TRADE SECRET
Map your electrical system and place a copy near the panel so that you can quickly identify an outlet later if a circuit breaker trips or a fuse blows.

DEVELOPING A FLOOR PLAN (CONTINUED)

Read up

Make a rough sketch of the work you propose, develop a rudimentary materials list, and then apply for a permit. At the time you apply, the building department clerk may be able to answer questions generated by the legwork you've done thus far. This feedback often proves invaluable.

Inspectors inspect

Inspectors are not on staff to tell you how to plan or execute a job, so make your questions as specific as possible. Present your rough sketch, discuss the materials you intend to use, and ask if there are specific requirements for the room(s) you'll be rewiring. For example, must bedroom receptacles have AFCI protection? Must kitchen wall receptacles be GFCIs if they are not over a counter? Be specific.

Draw up plans

Based on the feedback you've gotten, draw your plans. They should include each switch, receptacle, and fixture as well as the paths between switches and the device(s) they control. From this drawing, you can develop your materials list . Number each circuit or, better yet, assign a different color to each circuit. When you feel the plans are complete, schedule an appointment with an inspector to review them.

MATERIALS LIST: LIGHTING AND SWITCHES	

This materials list was determined from "Electrical Plan: Lighting and Switches Layer," on p. 169.

ROUGH

Fixture housings	Type 4, 6, and F1
4 ea.	3/0 or 4/0 metal boxes with bar hangars (verify fixture requirements)
3 ea.	4/0 NM boxes with bar hangars (wall fixtures only)
4 ea.	One-gang NM adjustable boxes
2 ea.	Two-gang NM adjustable boxes
4 ea.	Three-gang NM boxes
17 ea.	Romex connectors
500 ft.	14/2 NM Romex: 24 fixtures x 15 ft. ave. = 360 ft. (use excess for home runs, 3 x 50 = 150 ft.)
105 ft.	12/3 NM Romex (use remaining from power rough)

NOTE:
Staples, screws, and nail plates from power rough materials list.

Recommendations:
Verify all surface-mounted fixtures before rough. Some have very small canopies and must have a 3/0 or even a 1 gang box.

For undercabinet lights, do not install a box. Stub cable out of wall approximately 6 in. higher than the bottom of the upper cabinet. This way the drywall can be notched and the cable brought down to the perfect height by the cabinet installer. If undercabinet fixtures are to be installed at the front of the cabinet, some trim piece or metal sheath must be put over the NM cable to protect it from the wall to the fixture.

For recessed and other ceiling lighting, layout fixtures on the floor and use a plumb bob or laser to set fixtures on the ceiling.
To align straight rows of ceiling fixtures, use a taut line.

TRIM

Trims and lamps	Type 4, 6, F1
Fixtures and lamps	Type 1, 2, 3, 5, 8, 9,10
6 ea.	Single-pole Decora switches
5 ea.	Three-way Decora switches
2 ea.	Single-pole Decora dimmers
1 ea.	Four-way Decora dimmer
1 ea.	Decora timer
3 ea.	Three-gang plastic Decora plates
1 ea.	Two-gang Decora plate
4 ea.	One-gang plastic plates
4 ea.	Romex connectors

Listen well, take notes

Be low key and respectful when you meet with the inspector to review your plans. First, you're more likely to get your questions answered. Second, you'll begin to develop a personal rapport. Because one inspector will often track a project from start to finish, this is a person who can ease your way or make it much more difficult. So play it straight, ask questions, listen well, take notes, and—above all—don't argue or come in with an attitude.

On-site inspections

Once the inspector approves your plans, you can start working. In most cases, the inspector will visit your site when the wiring is roughed in and again when the wiring is finished. Don't call for an inspection until each stage is complete.

TRADE SECRET

If you're remodeling, keep in mind that every finish surface you drill or cut into is a surface that you'll have to patch later. So minimize cutting and drilling.

MATERIALS LIST: POWER

This materials list was determined from "Electrical Plan: The Power Layer," on p. 169.

ROUGH

18 ea.	One-gang NM adjustable boxes
1 ea.	One-gang metal cut-in box
1 ea.	Two-gang NM adjustable box (oven)
3 ea.	NM (Romex) connectors
1 ea.	Box (500) ¾-in. staples (to be used in rough lighting also)
1 ea.	Box (100) nail plates (to be used in rough lighting also)
500 ft	12/2 NM Romex (2 x 250ft. rolls): 20 units @ 15 ft. ave. = 300 ft (4 home runs @ 50 ft. ave. = 200 ft.)
250 ft.	12/3 NM Romex (1 x 250ft. roll): 2 home runs @ 50 ft. ave. = 100 ft. (use remainder in rough lighting)
50 ft.	10/3 NM (purchase cut to length): 1 home run @ 50 ft.
1	Bag (250) red wire connectors (to be used in rough lighting also)
1	Container (500) wafer-head #10 x ¾-in. screws

TRIM

16 ea.	Duplex receptacles 15-amp or 20-amp rated
3 ea.	GFCI receptacles (15 amp with 20-amp feed through)
1 ea.	30-amp/220v receptacle (verify with range manufacturer)
16 ea.	1 gang plastic duplex plates
1 ea.	30-amp/220v plate

NOTE:

GFCI's are packaged with their plates.

ELECTRICAL NOTATION

Start by making an accurate floor plan of the room or rooms to be rewired using ¼-in.-to-1-ft.-scale graph paper. Indicate walls and permanent fixtures such as countertops, kitchen islands, cabinets, and any large appliances. By photocopying this floor plan, you can quickly generate to-scale sketches of various wiring schemes.

Use the appropriate electrical symbols to indicate the locations of receptacles, switches, light fixtures, and appliances.

> ➡ **For a key of electrical symbols, see "Common Electrical Symbols," at right.**

Especially when drawing kitchens, which can be incredibly complex, use colored pencils to indicate different circuits. You can also number circuits, but colored circuits are distinguishable at a glance. Use solid lines to indicate cable runs between receptacles and switches and dotted lines to indicate the cables that run between switches and the light fixtures or receptacles they control.

The beauty of photocopies is that you can draw quickly. As you refine the drawings, refer back to the list of requirements given earlier to be sure that you have an adequate number of receptacles, that you have GFCI receptacles over kitchen counters, that there are switches near doorways, and so on. Ultimately, you'll need a final master drawing with everything on it. But you may also find it helpful to make individual drawings—say, one for lighting and one for receptacles—if the master drawing gets too busy to read.

If you have questions or want to highlight a fixture type, use callouts on the floor plan. As you decide which fixtures and devices you want to install, develop a separate materials list and use numbered keys to indicate where each piece goes on the master drawing. Finally, develop a list of all materials, so you'll also have enough boxes, cable connectors, wire connectors, staples, and so on. In short, list all you need to do the job.

> ➡ **For more on creating a materials list, see the charts on pp. 166-167.**

COMMON ELECTRICAL SYMBOLS

Symbol	
Duplex receptacle	
GFCI duplex receptacle	GFCI
Fourplex receptacle	
240v receptacle	
Weatherproof receptacle	WP
Switched receptacle	
Single-pole switch	S
Three-way switch	S_3
Switch leg	
Home run (to service panel)	
Recessed light fixture	ⓡ
Wall mounted fixture	
Ceiling outlet	
Ceiling pull switch	ⓢ
Junction box	ⓙ
Vent fan	VF
Ceiling fan	CF
Telephone outlet	
Two-wire cable	
Three-wire cable	

LIGHTING AND SWITCHES LAYER

Ⓐ Running 12/3 cable will accommodate two dedicated circuits in one cable, thus simplifying the wiring of the disposal and dishwasher.

Ⓑ Use 14/2 cable for all general lighting home runs (cable runs back to the service panel).

Ⓒ Verify the dimmer load; dimmers must be de-rated when ganged together.

POWER LAYER

Ⓐ Feed-through wiring of GFCI receptacle at beginning of run so it affords protection to receptacles downstream (see the top right drawing on p. 171)

Ⓑ Combo dishwasher/disposal receptacle. Install under sink in cabinet. Cut hot (gold) tab on receptacle to split receptacle for two circuits. Leave neutral (silver) tab intact. Be sure to install on two-pole breaker with handle tie.

Ⓒ Single-location GFCI protection. Do not feed through (see the bottom drawing on p. 171).

Ⓓ Stove is gas, so receptacle is only for igniter and clock and okay to run with hood. Leave NM cable stubbed at ceiling, leave 3 ft. to 4 ft. of slack for termination in hood/trim. (*NOTE:* Never run a stove igniter off GFCI-protected circuit as it will trip the GFCI every time the stove is turned on.)

Ⓔ Use 12/3 cable for home run, so a single cable takes care of dedicated refrigerator circuit and counter (small-appliance) circuit.

Ⓕ Home run for counter (small-appliance) circuit 2.

Ⓖ Oven outlet. Refer to unit specifications to verify receptacle or hard-wired connection.

ELECTRICAL PLAN: LIGHTING AND SWITCHES LAYER

A professionals' electrical floor plan may be daunting at first, but it'll start to make sense as you become familiar with the symbols used. To make the plans easier to read, they have been divided into two layers: (1) lighting and switches and (2) power, which consists of receptacles and dedicated circuits. (There's some overlap.) The *circled letters* are callouts that indicate areas warranting special attention. The *circled numbers* correspond to a list (on p.166) that specifies the type of fixture. Drawing switch legs and circuits in different colors makes a plan much easier to read.

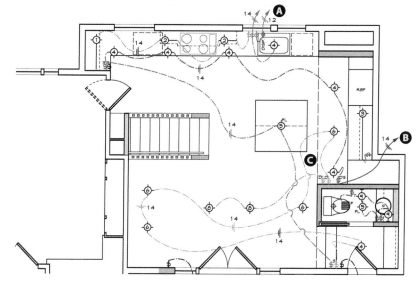

Drop lights in the dining area are noted by a dotted green switch leg running from the two ceiling boxes to the single-pole switch on the wall.

All circuits for recessed, drop, and undercabinet lighting are noted in different colors. Be certain to track each circuit back to the appropriate wall switch.

ELECTRICAL PLAN: THE POWER LAYER

This kitchen remodel is typical in that it has many dedicated circuits (also called designated circuits) and, per Code, at least two 20-amp appliance circuits wired with #12 cable. *Circled letters* are callouts that correspond to the lettered notes.

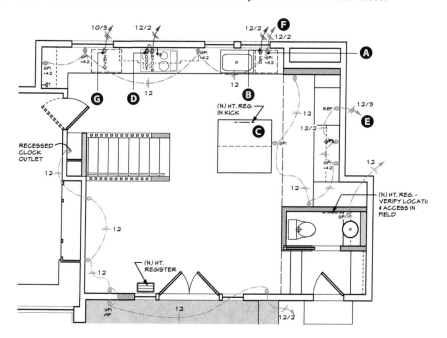

Whether for 110v wall outlets or 220v appliance outlets, each circuit should end with a home run back to the service panel.

Any receptacle that services the countertop must be GFCI protected. Refrigerators, however, should be run on a non-GFCI receptacle.

RECEPTACLES

RECEPTACLE IN MIDCIRCUIT

By splicing like wire groups and running *pigtails* (short wires) to the receptacle in this conventional method, you ensure continuous current downstream.

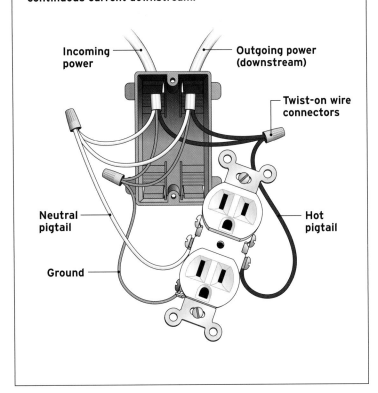

Incoming power

Outgoing power (downstream)

Twist-on wire connectors

Neutral pigtail

Hot pigtail

Ground

FEEDING WIRES THROUGH RECEPTACLE (not recommended)

Attaching hot and neutral wires directly to receptacle terminals is quicker and results in a less-crowded box. However, with this wiring method, if the receptacle fails, power can be disrupted to downstream outlets. *Note:* ground wires are *always* spliced to ensure continuity.

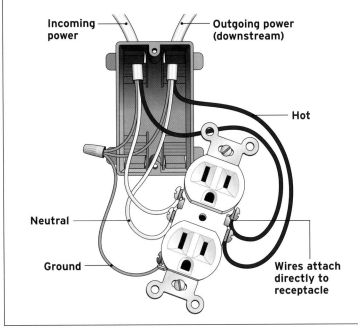

Incoming power

Outgoing power (downstream)

Hot

Neutral

Ground

Wires attach directly to receptacle

RECEPTACLE AT END OF CIRCUIT

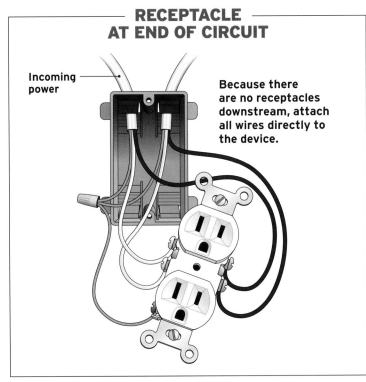

Incoming power

Because there are no receptacles downstream, attach all wires directly to the device.

The diagrams in this section show most of the circuit wiring variations that you're likely to need when wiring receptacles, fixtures, and switches. Unless otherwise noted, assume that incoming cable (from the power source) and all others are two-wire cable with ground, such as 14/2 w/grd or 12/2 w/grd (#14 wire should be protected by 15-amp breakers or fuses; #12 wire should be protected by 20-amp breakers). Although three-wire cable is common, it is used sparingly because it is significantly more expensive.

➡ For more wiring schematics, see Chapter 7.

All metal boxes must be grounded. Assume that non-grounded boxes in the wiring diagrams are nonmetallic (plastic) unless otherwise specified. Ground wires are indicated by green wires (although, in practice, ground wires in sheathed cables are typically bare copper). Black and red wires indicate hot conductors. (Some devices with multiple wire leads also use blue hot wires.) White wires indicate neutral conductors, unless *taped black* to indicate that the wire is being used as a hot conductor in a switch leg.

GANGED RECEPTACLES IN METAL BOX

A two-gang box with fourplex (*double duplex*) receptacles will be crowded. If the box is metal, use insulated ground pigtails and ground the box.

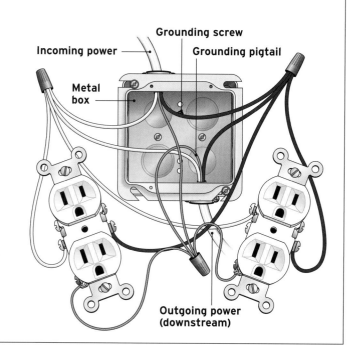

Grounding screw

Incoming power

Grounding pigtail

Metal box

Outgoing power (downstream)

GFCI RECEPTACLE, MULTIPLE-LOCATION PROTECTION

A GFCI receptacle can protect devices* downstream if wired as shown. Attach wires from the power source to terminals marked "line." Attach wires continuing downstream to terminals marked "load." As with any receptacle, attach hot wires to gold screws, white wires to silver screws, and a grounding pigtail to the ground screw. *Note:* Here, only ground wires are spliced; hot and neutral wires attach directly to screw terminals.

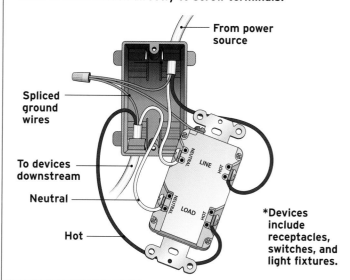

From power source

Spliced ground wires

To devices downstream

Neutral

Hot

NEUTRAL LINE HOT

NEUTRAL LOAD HOT

*Devices include receptacles, switches, and light fixtures.

GFCI RECEPTACLE, SINGLE-LOCATION PROTECTION

This configuration provides GFCI protection at one location—say, near a sink—while leaving devices downstream unprotected. Here, splice hot and neutral wires so the power downstream is continuous and attach pigtails to the GFCI's "line" screw terminals. With this setup, receptacle use downstream won't cause nuisance tripping of the GFCI receptacle.

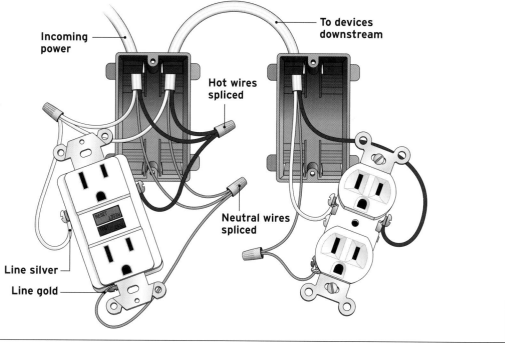

Incoming power

To devices downstream

Hot wires spliced

Neutral wires spliced

Line silver

Line gold

SWITCHES & LIGHTS

LIGHT FIXTURE AT END OF CABLE RUN

Switch wiring at its simplest: Incoming and outgoing hot wires attach to the terminals of a single-pole switch. Neutrals and ground are continuous.

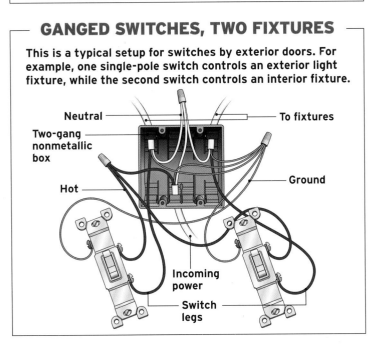

Grounding screw

Metal box

Ground

Nonmetallic box

Neutral

Incoming power

Hot

BACK-FED SWITCH AT END OF CABLE RUN

Attach the incoming neutral to a fixture lead; run the hot to a switch at the end of the cable. Use the white wire of a two-wire cable as one of the hot wires attaching to the switch—but *tape both ends of the white wire black* to show that it's hot.

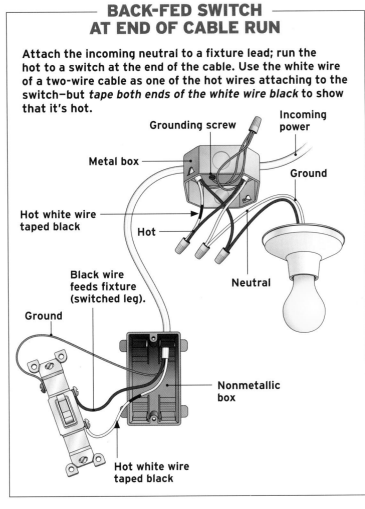

Grounding screw

Incoming power

Metal box

Ground

Hot white wire taped black

Hot

Black wire feeds fixture (switched leg).

Neutral

Ground

Nonmetallic box

Hot white wire taped black

GANGED SWITCHES, TWO FIXTURES

This is a typical setup for switches by exterior doors. For example, one single-pole switch controls an exterior light fixture, while the second switch controls an interior fixture.

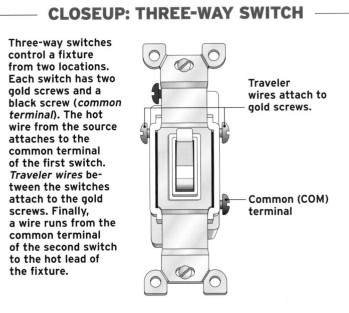

Neutral

To fixtures

Two-gang nonmetallic box

Hot

Ground

Incoming power

Switch legs

CLOSEUP: THREE-WAY SWITCH

Three-way switches control a fixture from two locations. Each switch has two gold screws and a black screw (*common terminal*). The hot wire from the source attaches to the common terminal of the first switch. *Traveler wires* between the switches attach to the gold screws. Finally, a wire runs from the common terminal of the second switch to the hot lead of the fixture.

Traveler wires attach to gold screws.

Common (COM) terminal

THREE-WAY SWITCHES, LIGHT FIXTURE BETWEEN

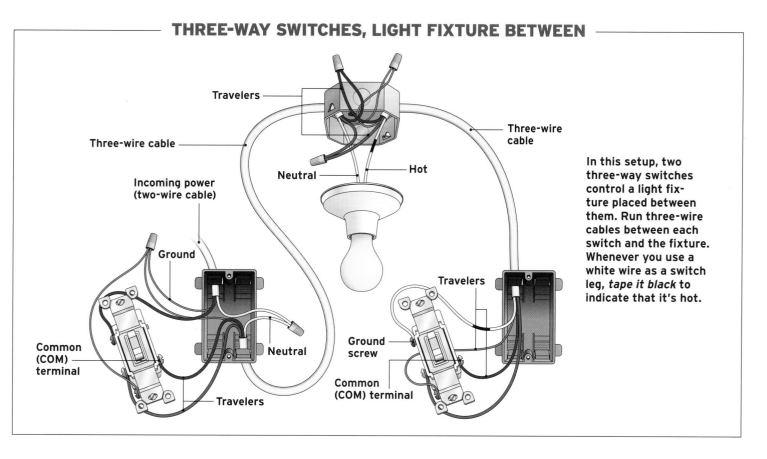

Travelers

Three-wire cable

Incoming power (two-wire cable)

Ground

Neutral

Hot

Common (COM) terminal

Neutral

Travelers

Three-wire cable

Travelers

Ground screw

Common (COM) terminal

In this setup, two three-way switches control a light fixture placed between them. Run three-wire cables between each switch and the fixture. Whenever you use a white wire as a switch leg, *tape it black* to indicate that it's hot.

THREE-WAY SWITCHES, LIGHT FIXTURE AT START OF CABLE RUN

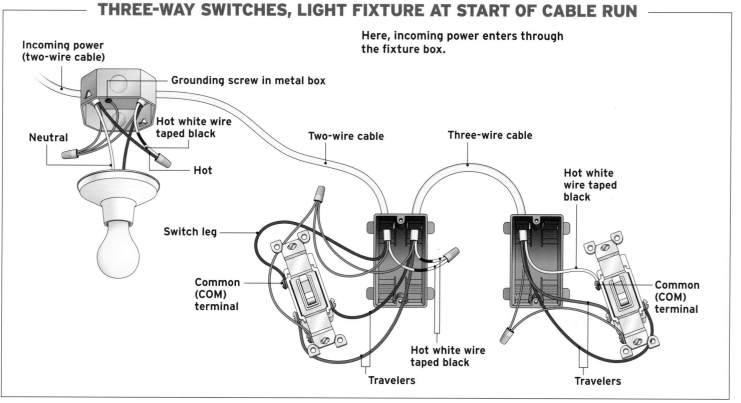

Here, incoming power enters through the fixture box.

Incoming power (two-wire cable)

Grounding screw in metal box

Hot white wire taped black

Neutral

Hot

Switch leg

Common (COM) terminal

Two-wire cable

Three-wire cable

Hot white wire taped black

Hot white wire taped black

Common (COM) terminal

Travelers

Travelers

RECEPTACLES, SWITCHES & LIGHTS

THREE-WAY SWITCHES, LIGHT FIXTURE AT END OF CABLE RUN

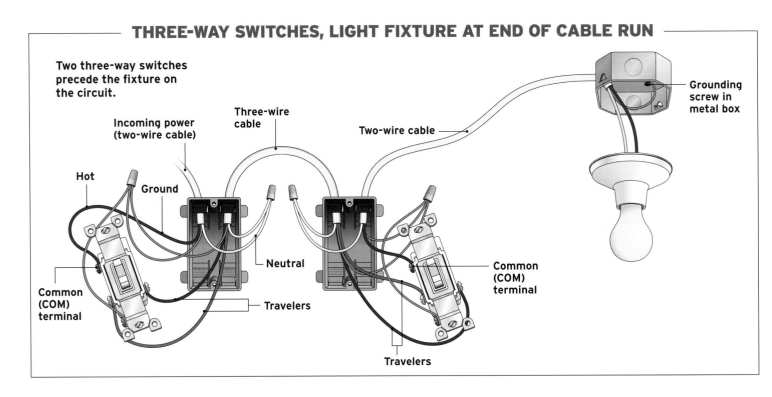

Two three-way switches precede the fixture on the circuit.

Incoming power (two-wire cable)

Three-wire cable

Two-wire cable

Grounding screw in metal box

Hot

Ground

Common (COM) terminal

Neutral

Travelers

Common (COM) terminal

Travelers

SPLIT-TAB RECEPTACLE CONTROLLED BY SWITCH, SWITCH UPSTREAM, REGULAR RECEPTACLE DOWNSTREAM

This setup is commonly used to meet NEC requirements if there is no switch-controlled ceiling fixture. As shown, the switch controls only the bottom half of the split-tab receptacle. The top half of the split-tab receptacle and all receptacles downstream are always hot. Removing the tab is shown in photo 1 on p. 49.

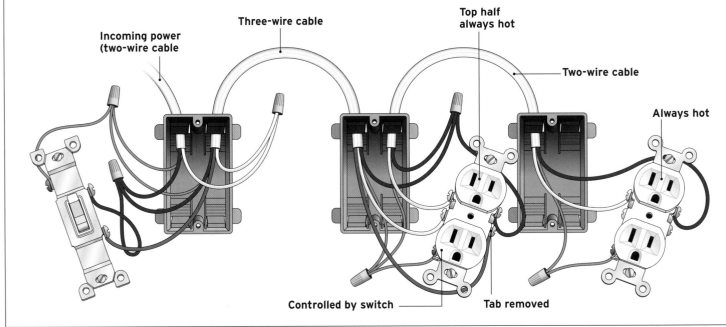

Incoming power (two-wire cable

Three-wire cable

Top half always hot

Two-wire cable

Always hot

Controlled by switch

Tab removed

SPLIT-TAB RECEPTACLE CONTROLLED BY SWITCH, SWITCH AT END OF CABLE RUN

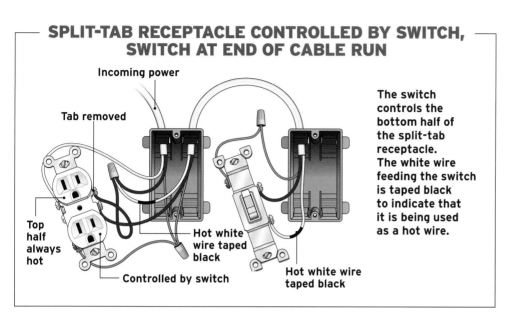

Incoming power

Tab removed

Top half always hot

Controlled by switch

Hot white wire taped black

Hot white wire taped black

The switch controls the bottom half of the split-tab receptacle. The white wire feeding the switch is taped black to indicate that it is being used as a hot wire.

SPLIT-TAB RECEPTACLE AT START OF CABLE RUN

The switch controls the bottom half of the split-tab receptacle on the left. The top half of the same receptacle stays hot, as does the regular duplex receptacle on the right. The white wire is taped black to show it is being used as a hot wire for a switch leg.

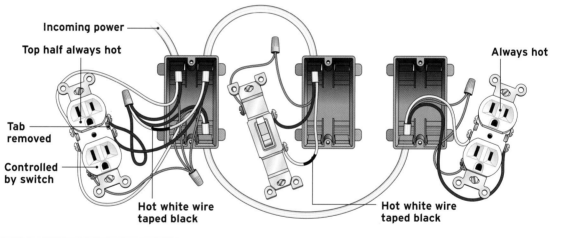

Incoming power

Top half always hot

Tab removed

Controlled by switch

Hot white wire taped black

Always hot

Hot white wire taped black

120/240V RANGE RECEPTACLE

This dedicated circuit requires an 10/3 cable with ground and a double-pole 30-amp breaker. Two 120v hot wires terminate to "hot" setscrews on the receptacle and the breaker poles; the neutral wire to the "neut" setscrew on the receptacle and the neutral/ground bus bar in the service panel.

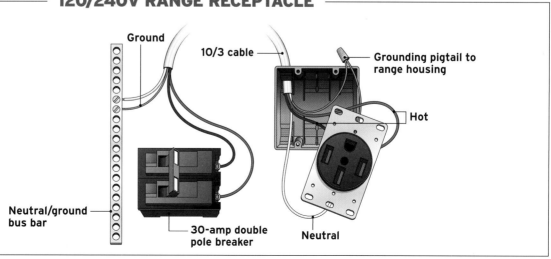

Ground

10/3 cable

Grounding pigtail to range housing

Hot

Neutral/ground bus bar

30-amp double pole breaker

Neutral

ROUGH-IN WIRING

ROUGH-IN WIRING REFERS TO THE first phase of a wiring installation. It is the stage at which you set outlet boxes and run electrical cable to them—as opposed to finish wiring, or connecting wires to devices.

Rough-in wiring is pretty straightforward when studs and joists are exposed. Whether a house is new or old, running wires through exposed framing is called *new work,* or new construction. If the framing is covered with finish surfaces such as plaster and drywall, however, the job is referred to as *remodel wiring.* Remodel wiring is almost always more complicated and costly because first you must drill through or cut into finish surfaces to install boxes and run cable and later you need to patch the holes you made.

ROUGH-IN BASICS

Wait until rough carpentry is complete before you begin rough-in wiring. Part of an electrician's job is setting boxes so they'll be flush to finished surfaces. Thus, before an electrician starts working, modifications to the framing—such as furring out or planing down irregular studs and ceilings joists—must be complete. If you're not sure how well the studs are aligned, hold a long straightedge across their edges and look for high and low spots.

Where to Start

Wait until the plumbers are gone. Waste pipes are large and often difficult to locate, which usually means a lot of drilling and cutting into studs and joists. Once the plumbing pipes are in place, you'll clearly see what obstacles you face and will have more room to move around.

Check your plans often. If there's not a table on-site where you can roll out your electrical plans, staple them to a stud—preferably at eye level so you can read them easily. Checking and rechecking the plans is particularly important if you're not a professional electrician.

Be flexible. As you lay out devices, you'll realize that not everything specified on the plans is possible; most plans are developed without knowing exactly what the framing looks like or where obstructions are. Be flexible and choose a solution that makes sense.

Organize Your Work

Perform one task at a time. Each task—such as setting boxes or drilling—requires a different set of tools. So once you have the tools out to do a given task, go around the room and complete all similar tasks. You'll become more proficient as you go: You waste less time changing tools, and the job goes much faster. In general, the sequence of rough-in tasks looks like this:

1. Walk the room with plans, noting outlet locations on walls and floor.
2. Snap chalklines or shoot laser lines to pinpoint box elevations and so on.
3. Attach boxes to studs and ceiling joists.
4. Drill holes for cable runs.
5. Pull cable through holes and into boxes.
6. Make up boxes—strip wire ends, splice wires, attach grounds, attach mud rings, and push wires into boxes.
7. Rough-in inspection.

After the inspection, finish surfaces are installed. Then, at the trim-out or finish stage, wires are attached to the devices.

The Rough-in Inspection

At the rough-in stage, inspectors look for a few key signs of a job well done: cables properly sized for the loads they'll carry; the requisite number and type of outlets specified by Code; cables protected by nail plates as needed; neat, consistent work throughout the system; and, above all, ground wires spliced and, in metal boxes, secured to a

When running cable around doors and windows, **find the easiest path. Consider drilling through wall plates and running cable in the floor above or below.**

During the rough-in inspection, **inspectors demand solid ground-wire splices and, in metal boxes, a ground screw or clip that secures the ground wire.**

ground screw. If grounds aren't complete, you won't pass the rough-in inspection.

At this inspection, only grounds need to be spliced. But since you've got the tools out, it makes sense to splice neutrals and continuous hot wires (those not attached to switches).

➡ **For more on grounding, see p. 12.**

When all splices are complete, carefully fold the wire groups into the box. When you come back to do the trim-out stage, simply pull the wires out of the box, connect wires to devices, and install devices and cover plates.

Connecting circuit wires to a main panel or subpanel is the very last step of an installation. As noted throughout this book, you should avoid handling energized cables or devices.

REMOVABLE WIRING SAFETY ESSENTIALS

Before removing box covers or handling wires, turn off the power to the area and use a voltage tester to be sure it's off.

First remove the fuse or flip off the circuit breaker controlling the circuit and post a sign on the main panel warning people of work in progress. Better yet, if you've got circuit breakers do as the pros do and install a breaker lockout so it will be impossible for anyone to turn it on. Breaker lockouts are available at electrical supply houses and most home centers.

➡ **For more on lockouts, see p. 238.**

Testing for power is particularly important in remodel wiring, because walls and ceilings often contain old cables that are energized. Here, an inductance tester is especially useful. Simply touch the tester tip to cable sheathing or wire insulation. That is, you don't have to touch the tester tip to bare wires to get a reading: If a cable, wire or electrical device is energized, the tip will glow. An inductance tester can detect current through cable sheathing.

Whatever tester you use, test it first on an outlet that you know is live to make sure the tester is working properly.

Always use a voltage tester **to test for power before touching cables, devices, or fixtures.**

⚠ WARNING

Make an emergency plan. It may be as simple as carrying a cell phone or having a friend close by—never do electrical work alone. Calling 911 is an obvious first step if an emergency occurs. On the job site, you should also post directions to the nearest hospital and a list of phone numbers of people to contact.

TOOLS FOR ROUGH-IN

Most of the tools you'll need for rough-in wiring are discussed at length in Chapter 2, so here we'll focus on tools that make rough-in easier and more productive.

Safety tools include voltage testers (especially an inductance tester), eye protection, work gloves, hard-soled shoes, kneepads, a dust mask, and a hard hat. Hard hats are clunky but essential when you're working in attics, basements or any other location with limited headroom. Every job site should also have a fire extinguisher and a first aid kit visibly stored in a central location.

Adequate lighting, whether drop lights or light stands, is essential to both job safety and accuracy. If a site is too dark to see what color wires you're working with, your chances of wrong connections increase.

Sturdy stepladders are a must. In the electrical industry, only fiberglass stepladders are Occupational Safety and Health Administration (OSHA) compliant because they're nonconductive. Wood ladders are usually nonconductive when dry, but if they get rained on or absorb ambient moisture, wood ladders can conduct electricity.

Shop vacuums, push brooms, dust pans, and garbage cans help keep the workplace clean. And a clean site is a safe site. Clean up whenever debris makes footing unsafe, especially if you're cutting into walls or ceilings. Plaster lath is especially dangerous because it's loaded with sharp little nails that can pierce the soles of your shoes.

Layout tools include tape measures, spirit levels, chalklines, and lasers. *Rough-in* is a misleading term because there's nothing rough or crude about locating fixtures or receptacles—layout is very exacting. For that reason, levels and plumb lasers are increasingly common when laying out kitchen and bathroom outlets **A**.

The most used hand tools are the same pliers, cutters, strippers, levels, and screwdrivers mentioned throughout this book. It's also handy to have a wire-nut driver **B** if you've got dozens of splices to make up.

There's also a nut-driver bit that fits into a screw gun **C**, but it tends to overtighten wire connectors if you're not experienced in using it. If you'll be cutting into and patching plaster or drywall, get a drywall saw and a taping knife. A flat bar (for prying) and an old, beat-up wood chisel are always busy in remodels, too.

Power tools help speed many tasks. Always wear eye protection when using them. A screw gun is preferable to nailing most of the time because you can easily remove a screw if you want to reposition, say, an outlet box.

A ½-in. right-angle drill is the workhorse of rough-in wiring because it has the muscle to drill through hard old lumber. If you use a standard 6-in.-long auger bit, the drill head and bit will fit between studs and joists spaced 16-in. on-center. It enables you to drill perpendicular to the framing to make wire pulling easier. Longer bits also have their merits: An 18-in. self-feeding auger bit **D** bores easily through several studs or doubled wall plates.

To cut larger holes in plaster or drywall—say, to retrofit pancake boxes or recessed light cans—use a fine-toothed hole saw.

A Lasers make for quick work when setting boxes.

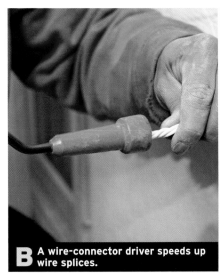

B A wire-connector driver speeds up wire splices.

D An 18-in. auger bit bores easily through several studs.

E A wire reel prevents kinking cable as you pull it.

C A nut-driver bit can overtwist spliced wires.

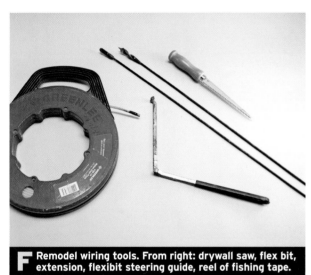

F Remodel wiring tools. From right: drywall saw, flex bit, extension, flexibit steering guide, reel of fishing tape.

A well-organized tool belt prevents you from looking for tools all day long.

Use a sabersaw to cut individual box openings into finish surfaces. When cutting through plaster lath, alternate cuts on each side of the opening rather than cutting one side completely. This will keep the lath from fluttering and cracking the plaster. Use a reciprocating saw to cut through framing or through plaster lath to create a cable trench.

➡ **For more on creating a cable trench, see p. 196.**

Use a demolition sawblade to cut through wood that contains nails or screws. It'll hold up to such heavy work.

There are a number of specialty tools designed to ease rough-in. A wire reel, a rotating dispenser that enables you to pull cable easily to distant points, is worth having. Reels hold 250 ft. of cable **E**. A 25-ft. fishing tape—a flexible steel, fiberglass, or nylon tape—enables you to pull cable behind finish surfaces **F**. In most cases, however, it's simpler to use a flexibit to drill through framing in one direction.

➡ **For more onpulling cable, see p. 197.**

A 48-in. drill extension will increase the effective drilling length of a flexibit. Use an insulated steering guide to keep the flexibit from bowing excessively. In a pinch, electricians wear a heavy work glove to guide a flexibit, but a glove may not protect you from shocks if you accidentally drill through an energized cable.

MATERIALS FOR ROUGH-IN

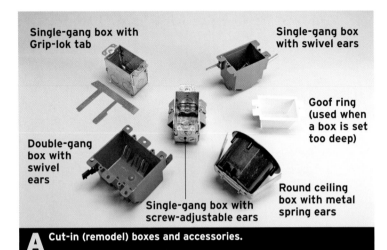

Single-gang box with Grip-lok tab

Single-gang box with swivel ears

Goof ring (used when a box is set too deep)

Double-gang box with swivel ears

Single-gang box with screw-adjustable ears

Round ceiling box with metal spring ears

A Cut-in (remodel) boxes and accessories.

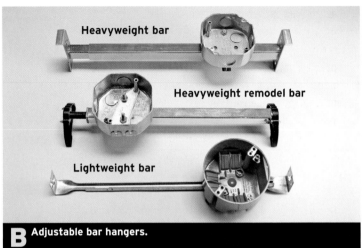

Heavyweight bar

Heavyweight remodel bar

Lightweight bar

B Adjustable bar hangers.

C A recessed light fixture with adjustable bar hangers.

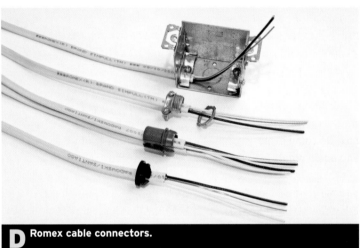

D Romex cable connectors.

As with tools, most of the materials installed during the rough-in phase were discussed in Section 2 and can be installed either in new construction or remodel wiring. There are, however, a number of specialized boxes, hanger bars, and other elements intended for remodel wiring that can be installed with minimal disruptions to existing finish surfaces.

Remodel (cut-in) boxes mount to existing finish surfaces—unlike new-work boxes, which attach to framing. Most cut-in boxes have small ears that rest on the plaster or drywall surface to keep boxes from falling into the wall or ceiling cavity. Spring clamps, folding tabs, or screw-adjustable wings on the box are then expanded to hold the box snug to the backside of the wall or ceiling. The devices that anchor boxes vary greatly **A**.

Code requires that exhaust fan boxes and ceiling boxes be mounted to framing. Expandable remodel bar hangers accommodate this requirement **B**, **C**.

Cable connectors (also called clamps) solidly connect cable to the box to prevent strain on electrical connections inside the box **D**. Cable clamps also prevent sheathing from being scraped or punctured by sharp box edges. Plastic boxes come with integral plastic spring clamps inside. If you use metal boxes, insert plastic push-in connectors into the box knockouts; no other cable connector is as quick or easy to install in tight spaces.

WHAT CAN GO WRONG

Don't forget to leave 12 in. to 18 in. of cable sticking out of each box for connecting devices later.

Ordering Materials

In general, order 10 percent extra of all boxes and cover plates (they crack easily) and the exact number of switches, receptacles, light fixtures, and other devices specified on the plans. It's okay to order one or two extra switches and receptacles, but because they're costly, pros try not order too many extras.

Cable is another matter altogether. Calculating the amount of cable can be tricky because there are infinite ways to route cable between two points. Electricians typically measure the running distances between several pairs of boxes to come up with an average length. They then use that average to calculate a total for each room. In new work, for example, boxes spaced 12 ft. apart (per Code) take 15 ft. to 20 ft. of cable to run about 2 ft. above the boxes and drop it down to each box. After you've calculated cable for the whole job, add 10 percent.

Cable for remodel jobs is tougher to calculate because it's impossible to know what obstructions hide behind finish surfaces. You may have to fish cable up to the top of wall plates, across an attic, and then down to each box. Do some exploring, measure that imaginary route and again create an average cable length to multiply. If it takes, say, 25 ft. for each pair of wall boxes and you have eight outlets to wire, then 8 outlets × 25 ft. = 200 ft. Add 10 percent, and your total is 220 ft. Because the average roll of wire sold at home center contains 250 ft., one roll should do it.

Have materials on hand when it's time to start installing boxes. Electricians often walk from room to room, dropping a box wherever floor plans indicate.

Rough-in recap: electrical code

- Circuit breakers, wiring, and devices must be correctly sized for the loads they carry. For example, 20-amp circuits require 12AWG wire and receptacles rated for 20 amps. Mismatching circuit elements can lead to house fires.

- All wire connections must be good mechanical connections. There must be good pressure between the connectors you are joining, whether wires are spliced together or connected to a device. For that reason, buy devices with screw terminals rather than back-wired (stab-in) devices whose internal clamps can deform. Deformed clamps can lead to loose wires, arcing, and house fires.

- All wire connections must be housed in a covered box.

- Boxes must be securely attached to framing so that normal use will not loosen them.

- Box edges must be flush to finish surfaces. In noncombustible surfaces (drywall, plaster) there may be a $1/4$-in. gap between the box edge and the surface. But in combustible surfaces, such as wood paneling, there must be no gap.

- All newly installed devices must be grounded. Code allows you to replace an existing two-prong receptacle or to replace a nongrounded box that has become damaged. However, if you install a new three-prong receptacle, it must be grounded. The only exception: You can install a three-prong receptacle into an ungrounded box if that new device is a ground-fault circuit interrupter (GFCI) receptacle. If you extend a circuit, the entire circuit must be upgraded to current Code.

- In new rough-in work, cable must be supported within 12 in. of any box and every $4^1/2$ ft. thereafter.

LAYING OUT THE JOB

With the electrical plans in hand, walk each room and mark box locations for receptacles, switches, and light fixtures. Each device must be mounted to a box which houses its wiring connections. The only exceptions are devices that come with an integral box, such as bath fans, recessed light cans, and undercabinet light fixtures **❶**.

Mark receptacles and switches on the walls first. Then mark ceiling fixtures. If studs and joists are exposed, use a vividly colored crayon that will show up on the wood. If there are finish surfaces, use a pencil to mark walls at a height where you can see the notations easily—these marks will be painted over later. Near each switch box, draw a letter or number to indicate which fixture the switch controls.

Once you've roughly located boxes on the walls, use a laser level **❷** to set exact box heights for each type of box.

1 Begin the lay out by walking the room and marking box locations.

2 Use a laser level to set box heights.

➡ **For more on box heights, see "Rough-in Recap: Box Heights," on the facing page.**

Use the laser to indicate the top, bottom, or center lines of the boxes **❸**. Many electricians prefer to determine level with the laser, then snap a permanent chalkline at that height so they can move the laser to other rooms **❹**.

To locate ceiling fixtures, mark them on the floor **❺** and use a plumb laser **❻** to transfer that mark up to the ceiling **❼**. This may seem counterintuitive, but it will save you a lot of time. Floors are flat, almost always the same size and shape as the ceilings above and—perhaps most important—accessible and easy to mark. In complex rooms, such as kitchens, draw cabinet and island outlines onto the floor as well. Those outlines will help you fine-tune ceiling light positions to optimally illuminate work areas.

3 Here, the laser beam indicates the centerline of the box.

4 You can snap a level chalkline in addition to using a laser.

Avoiding hot wires In remodels, there may be live wires behind finish surfaces. Use an inductance tester to test receptacles, switches, fixtures, and any visible wires. The cables feeding those devices will be nearby. Wall receptacles are usually fed by cables running 1 ft. to 3 ft. above. Switches often have cable runs up to a top plate; each ceiling fixture has cable running to the switch(es) controlling it. Avoid drilling or cutting into those areas, and you'll minimize the risk of shock.

5 Many electricians prefer to mark ceiling fixture centers on the floor.

6 They then use a plumb laser to transfer marks to the ceiling.

7 Center your hole saw on each ceiling mark.

WHAT CAN GO WRONG

When marking box locations on finish surfaces, use a pencil—never a crayon, grease pencil, or a felt-tipped marker. Pencil marks won't show through new paint. Also, grease pencils and crayons can prevent paint from sticking.

Rough-in recap: box locations

- Whatever heights you choose to set outlets and switches, be consistent.
- Code requires that no point along a wall may be more than 6 ft. from an outlet. Set the bottom of wall outlets 12 in. to 15 in. above the subfloor—or 18 in. above the subfloor to satisfy Americans with Disabilities Act (ADA) requirements.
- Place the top of switch boxes at 48 in., and they will line up with drywall seams (if sheets run horizontally), thus reducing the drywall cuts you must make.
- In kitchens and bathrooms, place the bottom of countertop receptacles 42 in. above the subfloor. This height ensures that each receptacle will clear the combined height of a standard countertop (36 in.) and the height of a backsplash (4 in.), with 2 in. extra to accommodate cover plates.

INSTALLING WALL BOXES

1 To install an adjustable box, press its bracket flush against the stud edge and screw the bracket to the edge or side of the stud.

2 Nail-on boxes cannot be adjusted, so use the depth gauge on the side to ensure that box edges will be flush to finish surfaces.

3 Boxes with integral brackets have small points at top and bottom that sink into the stud. A mud ring brings the box flush to the wall.

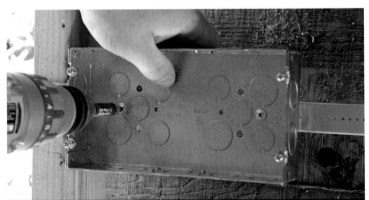

4 To provide solid support for wide multigang boxes, first install an adjustable box bracket that spans the studs.

Once you've established the heights of the switch and outlet boxes, installing them is pretty straightforward. Local electrical codes will dictate box capacity and composition.

➤ **For more on box capacity and composition requirements, see p. 25.**

In residences, 18-cu.-in. single-gang polyvinyl chloride (PVC) plastic boxes are by far the most commonly used. This size is large enough for a single outlet or a single switch and two cables. Otherwise, use a 22.5-cu.-in. single-gang box or a four-square box with a plaster ring.

Set each box to the correct height, then set its depth so that the box edge will be flush to the finish surface. If you use adjustable boxes, simply screw them to a stud **1**. To raise or lower the box depth, turn the adjusting screw. Side-nailing boxes typically have scales (gradated depth gauges) on the side **2**. If not, use a scrap of finish material (such as 1/2-in. drywall) as a depth gauge. Metal boxes frequently have brackets that mount the box flush to a stud edge **3**; after the box is wired, add a mud ring (plaster ring) to bring the box flush to the finish surface.

Multiple-gang boxes mount to studs in the same way. But if plans offset the box away from studs or a multigang box is particularly wide, nail blocking between the studs or install an adjustable box bracket and screw the box to it **4**. (The bracket is also called a *screw gun bracket,* because a screw gun is required to mount it and to attach boxes to it.) The more securely a box is supported, the more secure the electrical connections will be.

 TRADE SECRET
Most electricians prefer to screw boxes than to nail them because screws allow them to reposition boxes easily.

INSTALLING CEILING BOXES

INSTALLING A RECESSED CAN

1 When the fixture is where you want it, fasten its hanger bars to the joists.

2 Slide the can to fine-tune its position. Tighten the screws to lock it in place.

INSTALLING A BOX TO THE JOIST

3 If a box position needs to be offset slightly from a joist, nail it to blocking.

INSTALLING A BOX BETWEEN JOISTS

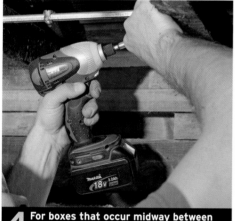

4 For boxes that occur midway between joists, use an adjustable bar hanger.

5 Screw a metal box to the hanger. Make sure the box edge is flush to the surface.

or

6 Alternately, you can nail 2x4 blocking between the joists.

B oxes for ceiling lights are most often 4-in. octagonal or round boxes or integral recessed light cans. Setting ceiling boxes in new work is similar to setting wall boxes, with the added concern that the ceiling box be strong enough to support the fixture weight. Many electricians prefer to use metal boxes for ceiling fixtures anyway. Ceiling fans require fan-rated boxes.

As noted earlier, you can mark ceiling box locations on the floor and use a laser to plumb up—or simply measure out from a wall.

➡ For more on laying out, see p. 184.

In most cases, you'll need to reposition the box to avoid obstacles or line it up to other fixtures, but it's quick work if the box has an adjustable bar hanger. To install a recessed can, for example, extend its two bar hangers to adjacent ceiling joists. Then screw or nail the hangers to the joists **1**. Slide the can along the hangers until its opening (the light well) is where you want it and then tighten the setscrews on the side **2**.

To install a 4-in. box, simply nail or screw it to the side of a joist. If you need to install it slightly away from a joist, first nail 2x blocking to the joist, then attach the box to

the blocking **3**. Remember: The box edge must be flush to finish surfaces.

To install a 4-in. box between joists, first screw an adjustable hanger bar to the joists **4**, then attach the box to it **5**. Alternately, you can insert 2x blocking between the joists and screw the box to it **6**.

➡ For a detailed look at hanger bars, see photo 2 on p. 85.

DRILLING FOR CABLE

Once boxes are in place, you're ready to run cable to each of them. It's rather like connecting dots with a pencil line. To prevent screws or nails from puncturing cables, drill in the middle of studs or joists whenever possible. If the holes you drill are less than 1¼ in. from the edge of framing members, you must install steel nail-protection plates. Always wear eye protection when drilling.

Drill for cables running horizontally (through studs) first. It doesn't matter whether you start drilling at the outlet box closest to the panel—called the *home run*—or at the last outlet on the circuit. Just be methodical: Drill holes in one direction as you go from box to box. However, if you're drilling for an appliance that has a dedicated circuit—and thus only one outlet—it's usually less work to drill a hole through a top or bottom plate and then run cable through the attic or basement instead of drilling through numerous studs to reach the outlet.

If possible, drill holes thigh high ❶. Partially rest the drill on your thigh so your arms won't get as tired. This method also helps you drill holes that are roughly the same height—making cable-pulling much easier. Moreover, when you drill about 1 ft. above a box, you have enough room to bend the cable and staple it near the box without crimping the cable and damaging its insulation.

For most drilling, use a 6-in., ³⁄₄-in.-diameter bit. Use an 18-in. bit to drill lumber nailed together around windows, doorways, and the like ❷. Using an 18-in. bit is also safer because it enables you to drill through top plates without standing on a ladder. Bits that bind suddenly can throw you off a ladder ❸.

TRADE SECRET

Drilled holes don't need to be perfectly aligned, but the closer they line up, the easier it is to pull cable. Some electricians use a laser to line up drill holes.

1 When drilling for cable runs, rest the drill on your thigh. This method eases drilling and places holes at a convenient height.

2 Use an 18-in. ship bit to drill through multiple studs.

3 The 18-in. ship bit also allows you to drill through top wall plates while standing on the floor.

PULLING CABLE

For greatest efficiency, install cable in two steps: (1) Pull cable between outlets, leaving roughly 1 ft. extra beyond each box for future splices and (2) then retrace your steps, stapling cable to framing and installing nail-protection plates. As with drilling, it doesn't matter whether you start pulling cable from the home run (first) box of a circuit or from the last box. If there are several circuits in a room, start at one end and proceed along each circuit, pulling cable until all the boxes are wired. Don't jump around: you may become confused and miss a box.

In new construction, electricians usually place several wire reels by the panel and pull cables from there toward the home run box of each circuit ❶. Once they've run cable to all the home run boxes, they move a reel next to each box and continue to pull cable outward until they reach the last (farthest) box ❷. When doing remodel wiring in a house with a crawl space, however, electricians often start at the last box and pull cables toward the panel. When they reach the home run boxes of several circuits, they will move the wire reels to those locations. From there, they feed, say, three cables down to a helper in the crawl space. The helper can pull all the cables toward the panel at the same time. This method is much faster than pulling single cables three different times.

Staple cable along stud centers to prevent nail or screw punctures. It's acceptable to stack two cables under one staple ❸, but use standoffs ❹ to fasten three or more cables traveling along the same path. (Multigang boxes are fed by multiple cables, for example.) Standoffs and ties bundle cables loosely to prevent heat buildup. As you secure cable, install nail plates where needed ❺.

1 Place a wire wheel near the beginning or end of each circuit.

2 Avoid crimping cable as you pull it through narrow spaces.

3 Drive staples just snug to the cable.

4 When bundling two or more cables, use standoffs.

5 For cable within 1¼ in. of a stud edge, install steel plate.

TRADE SECRET

Don't forget the home run. Put a piece of red tape on each home run box to ensure that you run cable from it to the panel. On a complex job with many circuits, you might run cable *between* all the outlets in a circuit but forget to install the home run cable that will energize the circuit. Not something you want to discover after the drywall's up.

FEEDING CABLE THROUGH CORNERS

1 To feed cable through a corner, drill a level hole on one side of the built-up studs.

2 Drill a second hole, at the same height, so the holes intersect.

3 Use diagonal cutters to cut through the sheathing and two of the three wires in the cable, then loop the end of the remaining wire.

4 Wrap the end of the cable so it slides through the hole easier, then work the looped wire through the holes.

Corners are usually built by nailing three or four studs together, so feeding cable through them can be tricky. Drill intersecting holes at the same height, as shown in photos ❶ and ❷. Use diagonal cutters to cut through the sheathing and two of the wires, thus leaving a single wire protruding from the cable ❸.

Use pliers to loop the end of the wire. Then wrap tape around the end of the cable and onto the wire to create a tapering point that will slide easier through the hole. Push wire through the corner holes until it emerges from the other side ❹. If the wire gets hung up midway through, insert a finger from the other side to fish for its end.

FISHING CABLE BEHIND FINISH WALLS

Most electricians hate fishing wire behind existing walls. It can be tricky to find the cable and time consuming to patch the holes in plaster or drywall. If you're adding a box or two, try fishing cable behind the wall. But if you're rewiring an entire room, it's probably faster to cut a "wiring trench" in the wall. Before cutting into or drilling through a wall, however, turn off power to the area.

➡ **For more on cutting a wiring trench, see p. 196.**

If you're adding an outlet over an unfinished basement, fishing cable can be straightforward. Outline and cut out an opening for the new box, insert a flexibit into the opening, then drill down through the wall's bottom plate ❶. When the bit emerges into the basement, a helper can insert one wire of the new cable into the small "fish hole" near the bit's point. As you slowly back the bit out of the box opening, you pull new cable into it ❷. No fish tape required! The only downside is that the reversing drill can twist the cable. This problem is easily avoided by sliding a swivel kellum over the cable end instead of inserting a cable wire into the flexbit hole ❸. Because the kellum turns, the cable doesn't.

Alternatively, you can start by removing a wall box. The closest power source is often an existing outlet. Cut power to that outlet and test to make sure it's off. The easiest way to access the cable is to disconnect the wires to the receptacle ❹ and remove it. Then remove the box ❺, which may require using a metal-cutting reciprocating-saw blade to cut through the nails holding the box to the stud. Fish a new cable leg to the location and insert the new and old cables into a new *cut-in box*. Secure the cut-in box to the finish surface, splice the cables inside the box, and connect pigtails from the splice to the new receptacle. >> >> >>

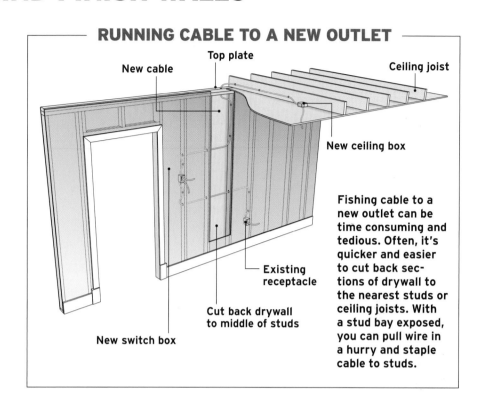

RUNNING CABLE TO A NEW OUTLET

New cable
Top plate
Ceiling joist
New ceiling box
Existing receptacle
Cut back drywall to middle of studs
New switch box

Fishing cable to a new outlet can be time consuming and tedious. Often, it's quicker and easier to cut back sections of drywall to the nearest studs or ceiling joists. With a stud bay exposed, you can pull wire in a hurry and staple cable to studs.

1 To fish cable to a new outlet, insert a flexibit into the new opening and drill.

2 Pull back after a helper below the floor attaches the cable.

FISHING CABLE BEHIND FINISH WALLS (CONTINUED)

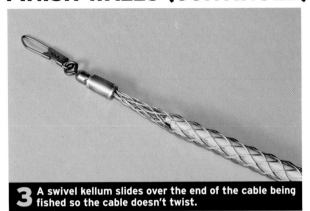

3 A swivel kellum slides over the end of the cable being fished so the cable doesn't twist.

or

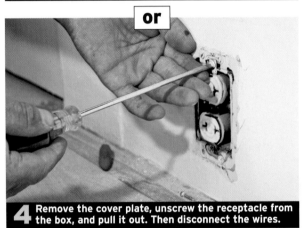

4 Remove the cover plate, unscrew the receptacle from the box, and pull it out. Then disconnect the wires.

5 Remove the old outlet box, fish new cable, and splice the old cable to the new. You'll replace the old box.

FISHING CABLE TO A CEILING FIXTURE

1 To run cable to a ceiling fixture, start by drilling an exploratory hole with a small-diameter bit.

Fishing cable to ceiling fixtures or wall switches is usually a bit complicated. If there is an unfinished attic above or a basement below, run the cable across it, then route the cable through a stud bay to the new box. To run cable to a ceiling light, for example, drill up through the fixture location using a 1/8-in. by 12-in. bit to minimize patching later ❶. Use a bit at least 6 in. long so a helper in the attic can see it—longer if the floor of the attic is covered with insulation. Measure the distance from the bit to the wall; a helper in the attic can use that measurement to locate the nearest stud bay to drill an access hole into. If you're working alone, loop the end of a stiff piece of wire about 1 ft. long ❷ and insert it in the drilled hole; friction will keep the wire upright in the hole until you can locate it in the attic.

If there's no access above the ceiling and/or cable must cross several ceiling joists to get from a switch to a light fixture, you'll have to cut into finish surfaces at several points ❸. To access cable in a stud bay, you'll need a cutout to expose the top plate. Using a flexibit may minimize the number of holes you must cut to drill across ceiling joists. But as noted earlier, it may ultimately take less time to cut and repair a single slot running across several joists than to patch a number of isolated holes. Whatever method you choose, make cuts cleanly to facilitate repairs. First outline all cuts with a utility knife.

2 If you're working alone, jam a long, looped wire into the exploratory hole, then go aloft to look for it.

3 Running cable from a ceiling fixture to a wall switch means cutting access holes.

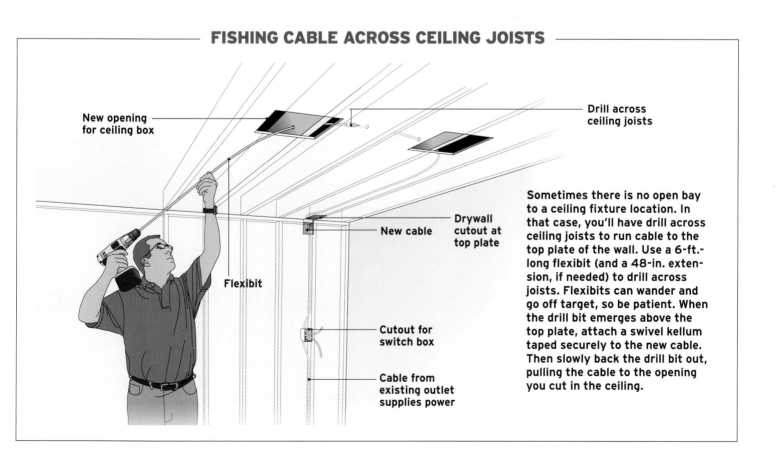

FISHING CABLE ACROSS CEILING JOISTS

New opening for ceiling box

Drill across ceiling joists

Flexibit

Drywall cutout at top plate

New cable

Cutout for switch box

Cable from existing outlet supplies power

Sometimes there is no open bay to a ceiling fixture location. In that case, you'll have drill across ceiling joists to run cable to the top plate of the wall. Use a 6-ft.-long flexibit (and a 48-in. extension, if needed) to drill across joists. Flexibits can wander and go off target, so be patient. When the drill bit emerges above the top plate, attach a swivel kellum taped securely to the new cable. Then slowly back the drill bit out, pulling the cable to the opening you cut in the ceiling.

CUTTING A WALL BOX INTO PLASTER

I f wiring is in good condition and an existing circuit has the capacity to add an outlet, turn off the power, cut a hole in the wall, fish cable to the location, and secure a cut-in box to the finish surface. This process is called *cutting-in* or retrofitting a box.

Hold the new box at the same height as other outlet or switch boxes and trace its outline onto the wall. Drill a small exploratory hole to locate studs or wood lath behind. If you hit a stud, move the box. If you hit lath, keep drilling small holes within the opening to find the edges of the lath. If you position the box correctly, you'll need to remove only one lath section ❶.

Use a utility knife to score along the outline to minimize plaster fractures. Remove the plaster within the outline using a chisel. Then cut out the lath, using a cordless jigsaw ❷. As you cut through the lath strip, alternate partial cuts from one side to the other to avoid cracking the plaster. Then carefully remove the plaster beneath the box ears, so they can rest on lath. Before inserting cut-in boxes, remove box knockouts, insert cable clamps, strip sheathing off the ends of incoming cable, and feed cable into the cable clamps. If more than one cable enters the box, write the destination of each on the sheathing. Secure the box by screwing its ears to the lath ❸.

1 Position the box and then trace the box outline onto the wall.

2 After you chisel out the plaster within the outline, use a sabersaw to cut lath.

3 After fishing and pulling cable, secure the box to the wall.

CUTTING A WALL BOX INTO DRYWALL

A dding a cut-in box to drywall is essentially the same as adding one to plaster. Start by drilling a small exploratory hole near the proposed box location to make sure there's no stud in the way.

There are a number of cut-in boxes to choose from; most common is a type with side-mounted ears that swing out or expand as you turn its screws.

➡ For more on box types, see p. 25.

Hold the box against the drywall, plumb one side, then trace the outline of the box onto the wall ❶. Drywall is much easier to cut than is plaster: Simply align the blade of a drywall saw to the line you want to cut and hit the handle of the saw with the heel of your hand.

There is no one right way to cut out the box, but pros tend to cut one of the long vertical sides, then make three short, horizontal cuts across. Then score and snap the last cut ❷. Finish the cutout with the drywall saw and a utility knife.

1 After drilling an exploratory hole and locating the box, trace the box outline onto the wall.

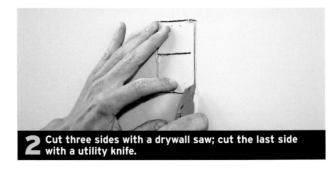

2 Cut three sides with a drywall saw; cut the last side with a utility knife.

RETROFITTING A CEILING BOX

As with all retrofits, turn off power to the area and explore first. Follow the mounting recommendations for your fixture. Attach the fixture box to framing.

If there's insulation in an attic above, remove it from the affected area. Be sure to wear eye protection and a dust mask when drilling through any ceiling—it's a dusty job.

Mark the box location and use a fine-tooth hole saw to cut through plaster or drywall ❶. Place the centering bit of the hole saw on the exact center of the box opening. Drill slowly so you don't damage adjoining surfaces—or fall off the ladder.

If the ceiling is drywall, you're ready to run cable through the hole in the ceiling. If the ceiling is plaster, cut through the lath or leave the lath intact and screw a pancake box through the lath and into the framing. Before attaching a pancake box, remove a knockout, test-fit the box in the hole, and trace the knockout hole onto the lath. Set the box aside, and drill through the lath, creating a hole through which you can run cable ❷.

Feed cable to the location and fit a cable connector into the box. Insert the cable into the connector, slide the box up to the ceiling ❸, and secure it ❹. Strip the cable sheathing and attach the ground wire to a ground screw in the box. Strip insulation from the wire ends and you're ready to connect the light fixture.

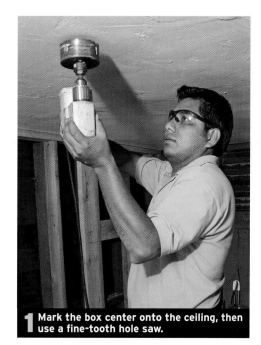

1 Mark the box center onto the ceiling, then use a fine-tooth hole saw.

2 Use an auger bit to drill a hole through the lath so you can pull cable to the box.

3 Fish cable through the lath and feed it through a connector in the box.

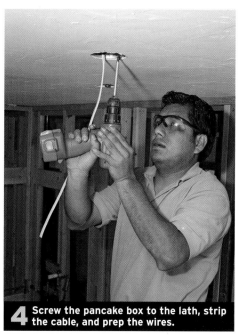

4 Screw the pancake box to the lath, strip the cable, and prep the wires.

TRADE SECRET
If you need to splice two or more cables, use a 4-in. octagonal or round box instead of a pancake box.

CREATING A WIRING TRENCH

1 Use a chalkline to mark the top and bottom lines of the wiring trench.

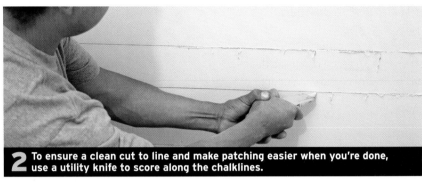

2 To ensure a clean cut to line and make patching easier when you're done, use a utility knife to score along the chalklines.

3 Cut along the chalklines, holding the reciprocating saw at a low angle. Do not cut the lath.

4 If walls are drywall, simply pull out the isolated strips. If plaster, use a hammer to remove the plaster and expose the lath.

5 Use a hammer and a flat bar to pry the lath free from the studs. Work slowly to minimize damage.

6 Pull out any nails still stuck in the studs; this will make later repairs easier.

When adding multiple outlets or rewiring an entire room, cutting a wiring trench in finish surfaces instead of fishing cable behind them is much faster. Before cutting or drilling, however, turn off the power to the areas affected. And be sure to wear eye protection and a dust mask.

If there are no windows in the walls to be rewired, cut the trench about 3 ft. above the floor so you won't have to kneel while working. If there are windows, cut the trench under the windows, leaving at least 1 in. of wall material under the windowsill to facilitate repairs. If there's plaster, make the trench as wide as two strips of lath. Snap parallel chalklines to indicate the width of the trench **1**. Then use a utility knife to score along each line **2**. Scoring lines first produces a cleaner cut and easier repairs.

Next use a reciprocating saw with a demolition blade to cut through the plaster or drywall. Hold the saw at a low angle: You'll be less likely to break blades or cut into studs **3**. Using a hammer, gently crush the plaster between the lines **4**. Use a utility bar (flat bar) to pry out the lath strips or drywall section. If you expose any cables in the walls, use an inductance tester to make sure they're not hot **5**. Next, drill through the studs so you can run cable in the trench. Wherever there's an outlet indicated, expand the trench width to accommodate the boxes.

Finally, walk along the trench and pull any lath or drywall nails still sticking in stud edges **6**. They're easy to overlook because they're small; if you pull them now, patching the trench will go smoothly.

RETROFITTING BOXES & PULLING CABLE

Once you've cut a wiring trench and drilled the holes, installing the boxes and pulling cable are straightforward and much like the basic sequences shown in Chapter 2.

➡ **For more on installing boxes, see p. 27.**

➡ **For more on pulling cables, see p. 191.**

Most electrical codes allow you to install either plastic or metal boxes in residences, but you must use metal boxes in commercial buildings.

If you're installing metal boxes, remove knockouts and insert cable connectors into their openings ❶. Then screw boxes to stud edges; screwing is less likely to damage nearby finish surfaces ❷. Be sure that the box will be flush to finish surfaces or, if you'll install plaster rings later, flush to the stud edge. Whenever you install boxes side by side—as with the outlet and low-voltage boxes shown here—install them plumb and at the same height ❸.

Installing cable in remodels can be tricky because space is tight and you must avoid bending cable sharply, which can damage wire insulation ❹. Install nail plates wherever the cable is less than 1¼ in. from the stud edges. Feed cable through the cable connectors into the boxes ❺. Finally, staple the cable to the framing within 12 in. of each box. If there's not room to loop the cable and staple it to a stud, it's acceptable to staple it to other solid framing, such as the underside of a sill ❻.

1 For metal boxes, start by removing knockouts and then insert a connector.

2 Screw the box to the side of a stud.

3 For the best appearance, install outlets that are side by side at the same height.

4 Run cable to each box.

5 Feed the cable through cable connectors until you have about 1 ft.

6 Staple the cable within 12 in. of each box.

MAKING UP AN OUTLET BOX

1 Use a cable ripper to slit sheathing, diagonal cutters to cut it.

2 A crimp tool may be used instead of a wire nut for the ground.

3 The long leg of the ground is looped beneath the grounding screw. This bonds

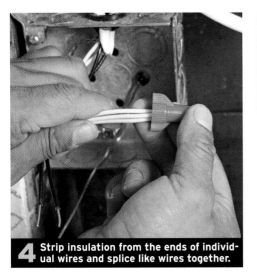

4 Strip insulation from the ends of individual wires and splice like wires together.

5 If there will be two receptacles in a box, make up two sets of pigtails.

6 Fold the wires into the box and install a plaster ring.

Electricians call the last stage of rough wiring *making up a box*. Here, you'll remove sheathing from cables inside the boxes, group like wires, splice grounds, and—if the box is metal—attach ground wires to a ground screw. At this point, many electricians simply splice all wire groups because it will save time later.

First use a cable ripper **1** to remove the cable sheathing. Leave a minimum of 1/4 in. of sheathing inside the box and a maximum of 1 in. Electricians favor utility knives to remove sheathing, but nonpros are less likely to nick wire insulation with a ripper. Once you've removed the sheathing, separate the insulated wires and bare ground wires.

There are several ways to splice grounds. You can use a twist-on wire connector (Wire-Nut is one brand) and run a pigtail to the ground screw. Or you can twist the bare wires and crimp them **2**.

If you go the latter route, leave only one ground wire sticking out of the crimp, which you'll wrap around the ground screw **3**.

Next, use a wire stripper to remove 1/2 in. of insulation from the wire ends. Use wire connectors to splice all neutral wires together and all hot wires together **4**. As we recommended earlier, splice a short pigtail to each wire group as well. Attaching neutral and hot pigtails to the receptacle screw terminals ensures continuity power even if a receptacle fails. If the box will contain two receptacles, create two groups of pigtails **5**.

Accordion-fold the wire groups into the box and you're ready for inspection. If necessary, install a mud ring to bring the box opening flush to finish surfaces. The mud ring should be applied only after you've repaired the plaster or drywall **6**.

MAKING UP A LIGHT CAN

Wiring a recessed light can or a ceiling box is essentially the same as making up an outlet box, although light cans frequently have stranded wire leads. Always follow the installation instructions supplied with your lighting unit.

Run cable to the light can, stapling it at least every 4½ ft. to the side of the ceiling joist and within 12 in. of the box. Also, make sure that the cable is at least 1¼ in. from the joist edge, so drywall screws or nails can't puncture it. Because you'll be working over your head and junction boxes are cramped, you may want to remove the sheathing before feeding cable into the box ❶. Tighten the cable connector to the small section of sheathing sticking into the box. Then strip insulation from individual wires ❷.

Group ground wires, neutral wires, and hot wires. Then splice each group using wire connectors. When wiring light cans, electricians typically start by splicing ground wires and running a ground pigtail to a ground screw on a metal box ❸. They then splice the neutral wires and, finally, the hot wires. When all wire groups are spliced, fold them carefully into the junction box and attach the cover.

1 Run enough cable to the light fixture so that there's about 1 ft. sticking out of its junction box.

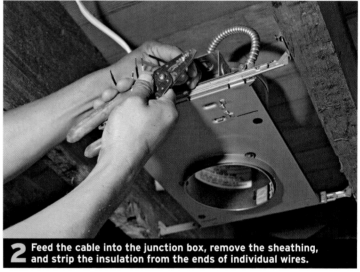

2 Feed the cable into the junction box, remove the sheathing, and strip the insulation from the ends of individual wires.

3 Use wire connectors to splice paired wire groups, starting with the ground wire group. Fold the spliced wires into the box.

TRADE SECRET
The ground screw must compress the ground wire evenly. Never cross the ground wire over itself because the screw would touch only that high spot.

MAKING UP A SINGLE SWITCH BOX

1 Score the sheathing lightly on both sides and give it a quick tug to slide it off the wires.

2 Use lineman's pliers to twist the grounds together. Leave one ground longer to be attached to the device.

3 Use a twist-on wire connector to splice the neutrals. Do not splice the hot wires; they are attached to the switch or receptacle later.

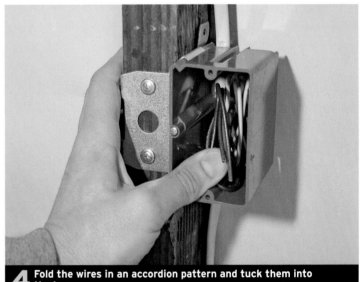

4 Fold the wires in an accordion pattern and tuck them into the box.

S witch boxes are most commonly plastic, so begin by using a screwdriver or needle-nose pliers to remove its knockouts. Feed cable (or cables) into the knockout and staple it within 8 in. of the box. There's no need to use cable connectors–plastic boxes have an integral spring clamp inside the knockout to prevent the cable from being yanked out. Leave about 1 ft. of extra cable sticking out of the box.

If there's a single cable entering the switch box, the switch is at the end of a switch leg and both insulated wires are hot wires. Remove the cable sheathing, strip insulation from the insulated wires, and wrap a piece of black electrical tape around the white wire to indicate that it is hot. Then tuck the wires into the box.

If there are two cables entering the switch box, one is incoming power and the other is a switch leg. Remove the cable sheathing ❶, then separate and twist the grounds together before splicing them with a wire connector ❷. Strip insulation from the ends of the neutral wires and splice them ❸. Fold the wires into the box ❹.

MAKING UP A MULTIGANG SWITCH BOX

The tricky part of making up a multigang switch box is keeping track of the wires. As you feed each cable into the box, use a felt-tipped marker to identify each cable's origin or destination. Write directly on the cable sheathing. Start by feeding the incoming cable into the box and marking it *110v*, *line*, or *hot*. This incoming cable will supply 110v from a panel or from an outlet upstream when the cable is connected later.

Next, nail a standoff to the side of the stud so you can secure each switch-leg cable as you feed it into the box ❶. Before inserting a cable into a knockout, however, note which fixture or device the switch leg controls ❷. Be specific and use labels such as *sconce*, *can ctr* (center can), and *perim* (perimeter fixture).

As you pull each cable into the box, strip the sheathing and cut off the small section that has writing on it ❸. When all the cables have been stripped, separate and twist the bare ground wires clockwise. Leave one ground about 6 in. longer than the others, then feed it through the hole in the end of a special ground wire connector ❹. This longer ground wire will run to a ground screw on each switch.

Strip the insulation from the ends of the neutral wires; then use a wire connector to splice them together ❺. Fold both ground- and neutral-wire groups into the box.

Cut a hot-wire pigtail 8 in. to 9 in. long for each switch and splice all pigtails to the hot wire of the incoming cable—the one that was earlier marked *110v* ❻. Finally, pair a hot pigtail with each switch leg ❼. Fold the wire pairs into the box. When it's time to wire the switches later, you'll connect a hot pigtail to one screw terminal and a switch-leg wire to the other screw terminal.

If a cable serves a three-way or four-way switch, indicate whether a wire is a "traveler" or "common."

➡ **For more on three- and four-way switches, see Chapter 3.**

1 Pull the incoming cable into the box, then nail a standoff to secure more cables.

2 Before pulling each cable into the box, indicate which fixture it controls.

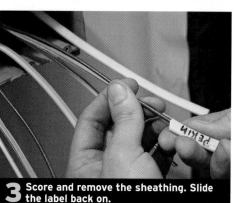

3 Score and remove the sheathing. Slide the label back on.

4 Feed the ground wire through the hole in the connector, then twist the connector.

5 Twist on a wire connector to splice the neutral wire group.

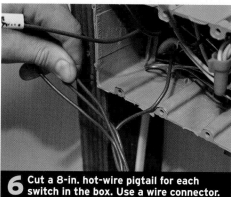

6 Cut a 8-in. hot-wire pigtail for each switch in the box. Use a wire connector.

7 Loosely wrap a hot pigtail around each switch-leg wire. Then fold into the box.

FLEXIBLE METAL CABLE

1 Flexible metal cable must be at least 1¼ in. from stud edges.

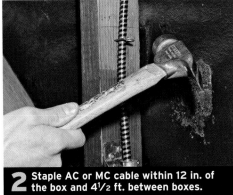

2 Staple AC or MC cable within 12 in. of the box and 4½ ft. between boxes.

3 To cut AC or MC cable, bend the cable till the jacket splits, then snip.

or

When using a Roto-Split, follow manufacturer's instructions.

CABLE CONNECTORS

Quick lock for MC

Snap-Tite® for MC

Locknut connector for AC

In residences, MC and AC cable is most often used in short exposed runs—typically between a wall box and the junction box of an appliance. The cable's metal jacket protects the wiring inside, but you should still take pains to avoid puncturing MC or AC cable with a nail or a screw. For this reason, when flexible metal cable is run through stud walls, it's a good practice to run it through the center of the stud **1**. If several cables run through a stud, stack holes vertically to avoid compromising the strength of the stud. Use nail-protection plates if the cable is closer than 1¼ in. to a stud edge.

➤ **For more on nail-protection plates, see p. 189.**

Secure or support flexible metal cable within 12 in. of boxes **2** and at least every 4½ ft. along the span of the run—electricians typically staple it every 3 ft. Flexible metal cable should also be secured to the underside of every joist it crosses or run through holes drilled through joists. Again, use nail-protection plates if the cable is less than 1¼ in. from a joist edge.

AC and MC cable require specialized connectors to secure them to boxes. You can use a hacksaw with a metal-cutting blade or diagonal cutters to cut through cable's metal jacket **3**, but a Roto-Split is the tool of choice. Whatever tool you use, cut through only one coil of the metal jacket to sever it. To prevent damaging the wire insulation within, cut no deeper than the thickness of the metal jacket. After cutting through the metal jacket, insert plastic antishort bushings to protect the wire insulation from the sharp edges of the metal jacket.

CABLE, CONDUIT, & MOISTURE

Moisture can short out electrical connections or, over time, degrade conductors and connectors. Thus it's important to use materials in appropriate locations.

- Nonmetallic (NM) sheathed cable (one brand is Romex): Dry locations only.

- Metal-clad (MC) or armored cable (AC): Dry locations; can be used in wet locations only if specifically listed for that use.

- Underground-feeder (UF) cable: Can be buried; when used in interiors, same rules as for NM cable.

- Electrical metallic tubing EMT): Wet or dry locations; okay in wet locations if fittings (couplings and connectors) are listed for use in wet locations; may be in direct contact with earth, if suitable and approved by local code.

- Rigid metal conduit (RMC): Same rules as for EMT; may be buried in earth or embedded in concrete.

- PVC plastic conduit (schedule-40 PVC): Can be exposed or buried underground; fittings are inherently raintight, so can be installed outdoors, in damp crawl spaces, and so on.

INSTALLING AC CABLE

To install AC cable, start by marking the box heights onto the studs and installing the metal boxes flush to finish surfaces or flush to the stud edges. Remove the box knockouts you'll need. Then drill the studs and run the AC cable through each hole. Staple it every 4½ ft. along the run and within 12 in. of each box.

Use a Roto-Split to cut through a single coil of the cable's metal jacket, then slide off the severed jacket section to expose the wires inside. Use diagonal cutters to snip off the kraft paper covering the wires ❶. Slide a plastic antishort bushing between the wires and the metal jacket ❷. Next, wrap the silver bonding wire around the outside of the jacket ❸. Wrapping this wire bonds the jacket and creates a continuous ground path.

Slide a setscrew connector over the end of the AC cable and tighten the setscrew to the metal jacket ❹. The screw compresses the cable jacket and the bonding wire, holding them fast and ensuring a continuous ground. Insert the threaded end of the connector into a box knockout. Then tighten the locknut that secures the connector ❺.

Attach a mud ring to the box to bring it flush to finish surfaces, which will be installed later, and fold the wires into the box ❻.

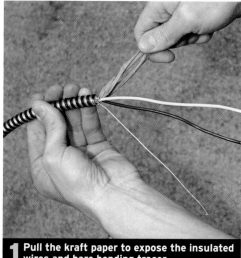

1 Pull the kraft paper to expose the insulated wires and bare bonding tracer.

2 Insert an antishort bushing between the wires and the jacket to prevent shorts.

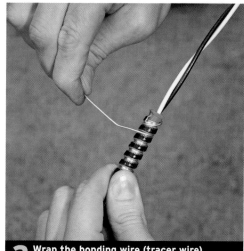

3 Wrap the bonding wire (tracer wire) around the outside of the jacket.

4 To ensure a solid ground path, slide a setscrew connector over the end of the cable.

5 To secure the AC cable to the box, tighten a locknut onto the connector.

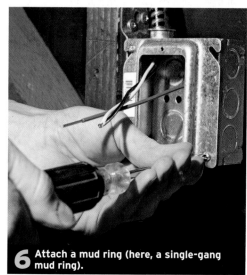

6 Attach a mud ring (here, a single-gang mud ring).

INSTALLING MC CABLE

MC cable is similar to AC cable, except that MC cable contains an insulated ground wire. Unlike AC cable, MC cable does not rely on its metal jacket to create a continuous ground for the circuit.

After using a hacksaw, diagonal cutters, or a Roto-Split to cut through the metal jacket of MC cable, slide on an antishort bushing to protect the wires inside from the sharp edges of the jacket ❶. Slide a quick-lock clamp over the end of the cable and snap the clamp into a box knockout. Then tighten the clamp screw to secure the MC cable ❷.

To create a continuous ground, use wire strippers to strip the ends of the incoming ground wire ❸. Use a wire connector to splice the incoming ground, the ground-screw pigtail, and a grounding pigtail that attaches to the device ❹. Then fold the wires into the box and screw on a mud ring to bring the box flush to finish surfaces, which are installed later.

TRADE SECRET

If you're pulling conduit wires in a residence, you probably don't need to use pulling lubricant. Use lubricant when:

- Pulling long distances—more than 100 ft.
- Pulling wires through multiple 90-degree bends.
- Pulling "fat" wires, such as 10AWG or larger.
- Pulling the maximum number of wires a conduit can hold.

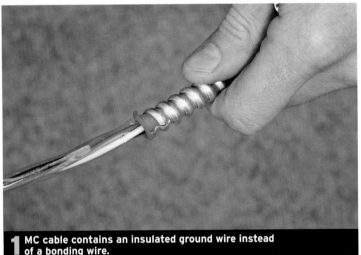

1 MC cable contains an insulated ground wire instead of a bonding wire.

2 MC cable does not rely on its metallic jacket to create a ground path. Thus you can use a quick-lock clamp to attach MC cable.

3 Strip the end of the incoming ground wire.

4 Use a wire connector to splice all three ground wires.

WORKING WITH EMT STEEL CONDUIT

1 To steady the pipe as you cut it, brace it with your legs.

2 Use a reaming tool to remove burrs, which could nick insulation.

3 Use a setscrew coupling to join lengths of conduit. Be sure to tighten the screws.

4 To secure pipe to a box, install a set-screw connector into a box knockout.

5 Use strap to secure conduit pipe.

TRADE SECRET

Use a hacksaw to cut metal conduit. Do not use the tubing cutters often used to cut plumbing pipe because they create a razor-sharp burr inside the conduit that is almost impossible to remove with a reaming tool.

Because EMT conduit is easy to work with, it's the rigid conduit type most commonly installed in residences. The main difference between conduit and flexible metal cable is that conduit comes without wire inside. Fortunately, pulling wire into conduit is a straightforward operation.

Assembling conduit is also straight forward. After cutting and deburring conduit pipe to length, insert the pipe ends into setscrew couplings and tighten the screws to secure the pipes. Threaded male fittings connect the pipe to boxes and Condulets®–covered fittings that facilitate wire-pulling or enable changes in pipe direction. You can buy prebent fittings such as capped-elbows (two-piece elbows that you can access to

pull wires) and sweeps (single-piece elbows with a wide turning radius). Or you can rent a conduit bender to change pipe direction.

By Code, EMT conduit must be strapped within 3 ft. of every junction box and no more than 10 ft. along runs. In actual practice, electricians strap pipe runs every 6 ft. to 8 ft. to prevent sagging.

Work from a layout sketch that indicates the power source, structural members, obstructions, existing outlets, and the locations of new outlets. Mount all the outlet boxes at the same height, and plumb each one. Plumbing boxes ensures that pipe connectors will line up with conduit pipes.

Use a hacksaw with a metal-cutting blade to cut EMT pipe; cuts need not be perfectly

square. EMT pipe typically comes in 10-ft. lengths ❶. After cutting the pipe, use a reaming tool to remove the burrs inside and out ❷.

To join EMT pipe, insert pipe ends into setscrew couplings ❸. To secure pipe to a box, use a connector on the box knockout ❹. In general, plumb vertical conduit sections before strapping them into place ❺. On the other hand, you may want pipe to follow an angled architectural element such as a brace.

MAKING TURNS WITH METAL CONDUIT

To change directions in a metal conduit system, you can either bend the pipe or install directional fittings such as offset adaptors, elbows, or Condulets. EMT pipe is rigid, but its walls are thin enough to bend easily with a conduit bender.

Pros bend conduit whenever possible. Bending pipe reduces the number of specialty fittings to buy and enables pipe to follow the contours of surfaces and structural elements. It's not necessary or desirable for conduit to follow every last jog or bulge in a wall; the simpler you can make an installation, the better it will look and the faster it will progress.

To bend pipe, use a felt-tipped marker to mark the beginning of the bend on the pipe. Slide the pipe into the bender ❶. Gently step on one side of the bender and simultaneously pull on the lever bar ❷. The raised marks on the outer curve of the bender indicate the angle you're creating in the pipe—typically, 15, 22½, 30, 45, or 60 degrees.

After bending the conduit—but before cutting it to final length—test-fit the piece to see if it lines up with the connector on the box ❸ or to a coupling that joins two pipe sections. With practice, you can also offset pipe ❹. Offsetting refers to creating two bends in opposite directions so a length of conduit can move from one plane to another.

There are a couple of rules to keep in mind as you bend conduit. First, there's a minimum requirement bend-radius for conduit: 10x the diameter. For ½-in.-diameter conduit, for example, the minimum bend radius is 5 in.

Second, each turn makes it harder to pull wire. So between each pair of boxes, you can have no more than 360 degrees of bends. In practice, every fourth turn should be a pulling point in which you can access and pull wire—in other words, the fourth turn should be a pulling elbow, a condulet, or a junction box. There is not enough room to splice wires in a pulling elbow or in a condulet. Splice wires only in a junction or outlet box.

1 Mark the beginning of the bend on the pipe; then position the pipe.

2 Pull and simultaneously step on one side of the bender.

3 Test-fit the piece to see if it lines up with the pipe connector.

4 Conduit benders can also create multiple offset angles.

USING DIRECTIONAL FITTINGS

You can also use directional fittings to make turns. A T-condulet enables you to run wires in different directions and doubles as a pulling point when fishing wire. To attach pipe to a condulet, first screw a locknut onto the threaded shaft of a male adaptor. Turn the adaptor most of the way into the Condulet hub and turn the locknut clockwise until it seats against the hub. Back-tightening the locknut in this manner ensures grounding continuity.

Use a T-condulet as a pulling point for fishing wires. Here, wires from the source diverge in two directions.

To attach pipe to a condulet, use an adapter and back-tighten the locknut until it lodges against the hub.

FISHING CABLE & CONDUIT

You can start fishing wire from either end of the circuit. If you're tapping into an existing outlet, it makes sense to fish from that outlet—after first turning off the power to the outlet and testing to make sure it's off.

In the installation shown here, a four-square extension box was mounted over an existing (recessed) box. Thus new wires can be pulled into the extension box and spliced to an existing cable to provide power for the circuit extension being added.

The fish tape can be fed easily into the conduit pipe ❶. At the other end of the conduit, tape wires to the fish tape ❷. To make wire pulling easier, leave the wire ends straight–do not bend them over the tape, but stagger them slightly so the bundled wires taper slightly. Wrap electrical tape tightly around the wire bundle so that wires stay together and won't snag as they're pulled. The pulling will also go easier if you pull stranded wire rather than solid wire, which is stiffer and less flexible.

Finally, use a wire caddy ❸, even if you have to build one out of scrap pipe and lumber. Using wire spools on a caddy helps minimize tangles.

TAPPING INTO AN EXISTING OUTLET

Tapping into an existing outlet is often a convenient way to supply power to a conduit extension. Remove the cover plate from the outlet, detach wires from the old receptacle, then use lineman's pliers to straighten the old wires so they'll be easier to splice to wires running to the new outlets.

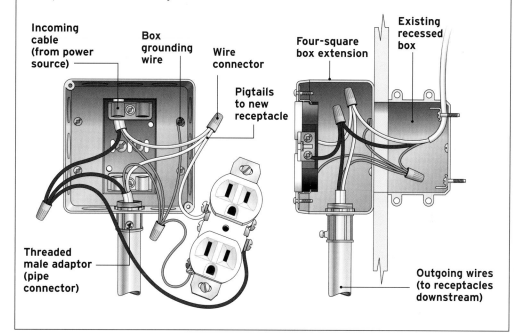

1 Push the fish tape down into the conduit.

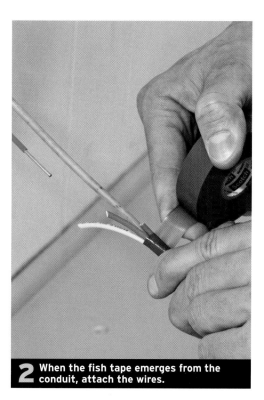

2 When the fish tape emerges from the conduit, attach the wires.

3 As you pull wires through the conduit, the spools on the wire caddy turn.

PREPPING RECEPTACLES FOR SURFACE METAL BOXES

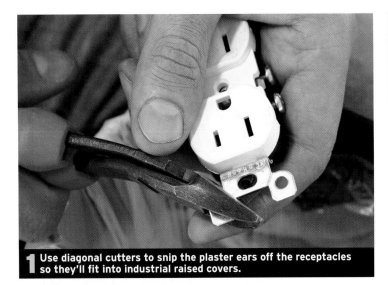

1 Use diagonal cutters to snip the plaster ears off the receptacles so they'll fit into industrial raised covers.

2 Attach solid pigtails to screw terminals.

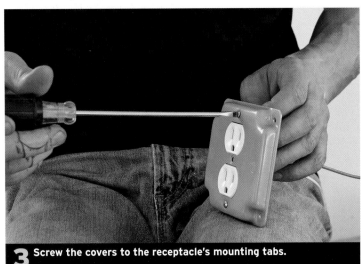

3 Screw the covers to the receptacle's mounting tabs.

4 Screw the ground pigtail to the box.

To save time, professionals often divide receptacle wiring of a conduit system into several smaller tasks, each of which can be done with a single tool. Your conduit system may have slightly different materials, but these prep techniques should save you time.

Standard duplex receptacles come with plaster ears intended to seat against plaster or drywall surfaces. Consequently, the ears may not fit into metal boxes or behind the industrial raised covers often installed in conduit systems. Use diagonal cutters to snip off the ears **1** so they fit inside such covers.

To speed the assembly, precut all the pigtails that you'll attach to receptacles. To loop pigtail ends, strip ½ in. of insulation, insert the stripped ends into the little hole on the stripper jaw, and twist your wrist.

Use a screw gun to attach the looped ends to the receptacle screw terminals **2**. Remember to loop the pigtail ends clockwise so that when the screw tightens (clockwise), the loop stays on the screw.

Next, attach the prewired receptacles to the covers. Typically, there is a pair of machine screws (and nuts) that screw into the mounting tabs at top and bottom, and a single short screw for the center hole in the front of the receptacle **3**.

Finally, screw a ground pigtail to the box **4**. The threaded hole for the ground screw is raised slightly to facilitate surface mounting. If you surface mounted a box on a concrete wall and the box didn't have this detail, the ground screw would hit concrete before it tightened all the way down—thus jeopardizing grounding continuity.

WIRING OUTLETS

1 Strip ½ in. of insulation from the ends of the stranded wire.

2 Splice the ground wires first.

3 After splicing like-colored wire groups, fold them into place.

4 Then screw the receptacle cover to the box.

GROUNDING CABLE & CONDUIT SYSTEMS

Every newly installed circuit must have continuous grounding. Steel conduit acts as its own ground path. MC cable contains an insulated ground because its metal jacket does *not* serve as a ground path; it exists solely to protect the wires inside. AC cable's metal jacket, on the other hand, *does* serve as a ground path. To ensure a continuous ground in AC cable runs, wrap and secure the cable's thin silver bonding wire.

If Code requires steel conduit, AC or MC, you must use steel boxes. Steel boxes must be grounded as well. In addition to tightening the cable or conduit couplings to the box knockouts, screw a grounding pigtail into the threaded hole in the box. The box pigtail is then spliced to circuit grounds and pigtails that run to the device.

Alternatively, if you're using solid wire, you can use a grounding clip to ground a metal box. (Code prohibits using a grounding clip with stranded wire.) If the incoming ground wire is insulated, strip approximately 6 in. of the insulation and slide the grounding clip onto a section of bare wire. This will allow enough wire beyond the clip so you can splice the bare wire end to other grounds or attach it directly to the grounding screw on a device.

Premade grounding pigtails come with one end looped around a green ground screw. If your job is large, using premade grounds can save a lot of time.

Using a ground clip to ground a metal box is suitable for solid connectors only.

If you've already attached pigtails to the receptacle screw terminals, the connections at outlet boxes on a conduit system will go quickly. *Note:* In the project shown here, the red wire is the hot conductor, the white wire is neutral, and green is the ground conductor. Code requires that green or bare wires are always the designated ground.

Use wire strippers to strip ½ in. of insulation from the wires pulled through the conduit **1**. Splice the receptacle pigtails to like-colored circuit wires pulled through the conduit. Typically, electricians splice the ground wires first, which means splicing three wires: the receptacle pigtail, the grounding pigtail to the box, and the circuit ground **2**. If the metal conduit serves as the ground, there will be no circuit ground wire.

When splicing stranded wire to solid wire, strip the stranded wire slightly longer than the solid wire, so the stranded-wire end sticks out beyond the solid wire. By doing this, you force the stranded-wire end into the wire connector first, ensuring a solid connection. Use lineman's pliers to twist wires slightly before screwing on the connector.

Once you've spliced all wire groups, fold the wires into the box **3** and attach the cover. Hold the cover tight against the box and screw in two diagonally opposite screws—there's no need to use four screws to attach a cover, although most have four screw holes **4**.

APPLIANCES

WIRING APPLIANCES, LIKE WIRING general-use circuits, is primarily a matter of solidly connecting conductors and following the recommendations of the NEC and the manufacturer. Many appliances are big energy users, so it's particularly important to size circuit wires and breakers based on appliance loads. When your new appliances arrive, read—and save—the owner's manuals that come with them. Owner's manuals contain essential information on how to install the appliance, how to register the appliance and comply with its warranty, how to identify and order replacement parts, and so on. Increasingly, manufacturers offer owner's manuals online, so if your appliance lacks a manual, download a copy at once. There's no guarantee that manuals will stay in print—or online.

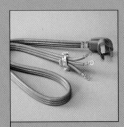

BASICS

GARBAGE DISPOSERS

OVENS

HOUSEHOLD APPLIANCES

A QUICK LOOK AT KITCHEN APPLIANCES

There's a lot going on in a modern kitchen. Typically it contains a refrigerator, a dishwasher, a garbage disposer, a range hood, a slew of small countertop appliances, and, in many cases, an electric range, cooktop, or oven.

➤ For more on wiring range hoods, see p. 152.

Appliances have different wiring requirements. Smaller ones such as disposers, dishwashers, and range fans require 120v; other appliances require 240v; and others—such as electric ranges—require both 120v and 240v. Ranges burners and ovens use 240v, but a range's timer and clock use 120v.

Let's start with a handful of important concepts.

Covered junction boxes

All electrical connections must take place in a covered junction box—either an integral box inside the appliance or in a covered junction box secured to a framing member, such as a wall stud.

Equipment grounding

All appliances must have an equipment-grounding conductor that connects to both the appliance frame (or housing) and to the metal outlet box for the safe discharge of fault currents.

Accessible disconnect means

Appliances typically have an accessible "disconnect means" so you can cut the power.

(This is in addition to the fuse or breaker controlling the circuit.) For appliances that slide out—such as a dishwasher or a refrigerator—the disconnect means is typically a cord and a receptacle plug. Appliances that are *hard wired,* such as drop-in cooktops and wall ovens, must have an access panel near the junction box where incoming circuit wires connect to *appliance whips.* (Whips are flexible cords—metal clad or plastic sheathed—that connect to appliance terminals.)

Dedicated circuits

Code requires that every fixed appliance must be served by a separate, *dedicated* circuit—a circuit that serves only that appliance. This includes appliances that must stay on, such as refrigerators and freezers, as well as large energy users, such as electric ranges, electric water heaters, and clothes dryers that require heavier wire and higher-rated breaker (or fuse) protection.

➤ For more on dedicated circuits, see p. 167.

COMMON ELECTRICAL APPLIANCE REQUIREMENTS		
User	Typical Wire Size*	Fuse or Breaker
Small-appliance circuit	12AWG	20 amps
Refrigerator (120v)[†,‡]	12AWG	15 amps
Stand-alone freezer (120v)[†,‡]	12AWG	15 amps
Dishwasher (120v)[†,‡]		
Disposer (120v)[†,‡]	12AWG	20 amps
Microwave (120v)[†,§]	12AWG	20 amps
Range (120/240v)[†,‡]	10AWG	30 amps
Range (120/240v)[†,‡]	8AWG	40 amps
Range (120/240v)[†,‡]	6AWG	50 amps
General utility/workshop	12AWG	20 amps
Laundry circuit (washer)[ᴾ]	12AWG	20 amps
Clothes dryer (120/240v)[†,‡]	10AWG	30 amps
Water heater (240v)[†,‡]	10AWG	30 amps
Baseboard heater (120v)[†,‡]	12AWG	20 amps (max. 1,500w on circuit)
Baseboard heater (240v)[†,‡]	10AWG	30 amps (max. 5,760w on circuit)
Whole-house fan[†]	12AWG	20 amps
Window air-conditioner (240v)[†,‡]	10AWG	20 amps

*Ratings given for copper (CU) wiring.
[†]Requirements vary; check rating on appliance nameplate and follow manufacturer's specifications.
[‡]Requires dedicated circuit.
[§]Microwaves are typically rated 15 amps but are installed on 20-amp kitchen circuits.
[ᴾ]Requires designated circuit (see p. 219).

TRADE SECRET
Don't use ground-fault circuit interrupter (GFCI) receptacles with major appliances, such as refrigerators. The motors of such appliances have a high inductance load as they start up, which can cause a GFCI receptacle to trip unnecessarily—leaving you with a refrigerator full of spoiled food.

PREPARING AN UNFINISHED APPLIANCE CORD

There are many different types of appliance cords. Some cords come with a molded plug and precrimped connectors that attach to terminals on the appliance. Other cords have a molded plug but an unfinished end that you must strip and splice to the lead wires of an appliance. In most cases, the splice is housed in an integral junction box inside an appliance.

The cord seen here is a typical 120v cord that you might attach to a smaller fixed appliance, such as a garbage disposer. It contains a hot wire, a neutral wire. and—in the center—a sheathed ground wire. Look closely at the cord and you'll see that its sheathing has a ribbed side and a smooth side. The ribbed side contains a neutral wire that must be spliced to the neutral wire of the appliance; the smooth side contains a hot wire ❶.

Start by snipping and separating the three stranded wires within the cord ❷. Using a utility knife, carefully slice and peel back the cord's outer (gray) insulation from the middle wire. As you do so, you'll expose the ground wire's green insulation. Only the ground wire has this additional layer of insulation ❸. Next use a wire stripper to remove ½ in. of insulation from the ends of all three cord wires ❹. Now you're ready to splice those wires to the lead wires on the appliance.

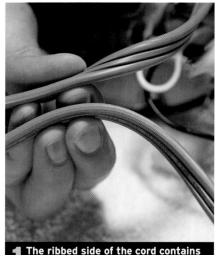

1 The ribbed side of the cord contains the neutral wire.

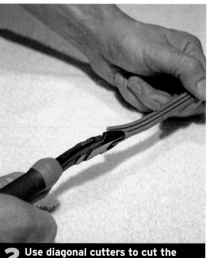

2 Use diagonal cutters to cut the cord and separate the three wires.

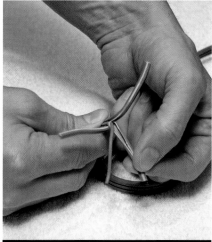

3 Peel back the gray insulation to expose the ground wire.

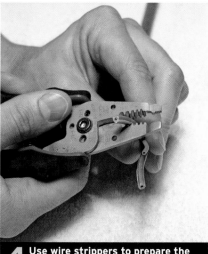

4 Use wire strippers to prepare the ends of all three cord wires.

WHIP IT!

For eons, how-to books had sections about replacing appliance plugs. Forget that. Today, you can easily find replacement cords—also called appliance whips—with molded plugs and precrimped connectors, which are far easier and safer to install. By the way, always grab the plug—not the cord—when unplugging an appliance.

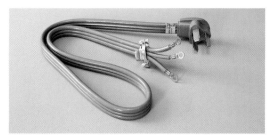

Replacing the entire cord is often easier than replacing a broken plug.

The connectors on the whip attach to terminals on the appliance.

INSTALLING A GARBAGE DISPOSER

1 The bundled yellow wires supply the garbage disposer and dishwasher.

2 Remove the cover plate from the junction box and screw on a connector.

3 Feed the cord wires through the connector and junction box.

4 Tighten the cable connector to prevent stress on the wire connections.

5 Attach the green ground wire to the grounding screw in the junction box.

6 Raise the disposer into place and turn it until the mounting rings lock into place.

Though installation of a disposer is largely the same from brand to brand, be sure to follow the instructions that come with your unit. Typically, plumbing supply and waste pipes are stubbed out, and 12AWG cable is roughed-in before the finish walls and cabinets are installed **1**. After the base cabinet is in place, install a 20-amp duplex receptacle to supply power to the disposer and dishwasher. The receptacle for the disposer should be a switched receptacle.

➤ **For more on wiring a split-tab receptacle, see p. 49.**

Install the sink and attach the disposer mounting assembly in the sink outlet. Then route the dishwasher overflow pipe into the cabinet under the sink.

With these prep steps done, you're ready to attach the appliance cord. Remove the cover plate from the bottom of the unit and pull its wire leads. Then screw a cable connector into the knockout in the bottom of the unit **2**. Separate and strip cord wires and feed them through the cable connector. Use needle-nose pliers to pull the wires through the junction box **3** and then tighten the cable **4**.

Attach the green ground wire to the ground screw in the junction box. Then use wire connectors to splice like wires—neutral to neutral, hot to hot **5**. Fold all wires into the junction box and replace the cover plate. Lift the disposer until its mounting ring engages the mounting ring on the bottom of the sink. Turn the unit until the rings lock **6**.

Slide the tube from the dishwasher onto the dishwasher inlet stub and tighten its clamps **7**. Attach the P-trap to the discharge outlet on the disposer **8**. Plug the disposer plug into the undersink receptacle. The second plug runs to the dishwasher.

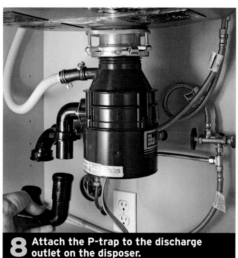

7 Slide the dishwasher drain onto the inlet stub and tighten its clamp.

8 Attach the P-trap to the discharge outlet on the disposer.

DISPOSERS & DISHWASHERS

Disposers and dishwashers are both 120v, 20-amp appliances, so both are supplied by 12AWG cable. Most garbage disposers have a covered junction box on the bottom, to which a plugged cord attaches. You can also hard wire a disposer, but having a receptacle in the cabinet under the sink enables a homeowner to quickly unplug the unit should he or she need to repair or replace it. Typically, an undersink outlet is controlled by a switch above the counter, as shown in "Wiring a Garbage Disposer," at left.

A dishwasher and a garbage disposer usually plug into a duplex receptacle in the cabinet under the sink—a split-tab receptacle. A split-tab receptacle is a standard duplex receptacle whose middle tab has been removed to create a duplex receptacle fed by two circuits—that is, by two hot wires.

Because they slide out for installation and maintenance, dishwashers are also installed with a cord and plug. Most of the time, the dishwasher junction box is located in the front of the unit, just behind the kick panel. After attaching cord wires to wire leads in the junction box, run the cord in the channel behind the dishwasher to reach to an outlet.

WIRING A GARBAGE DISPOSER

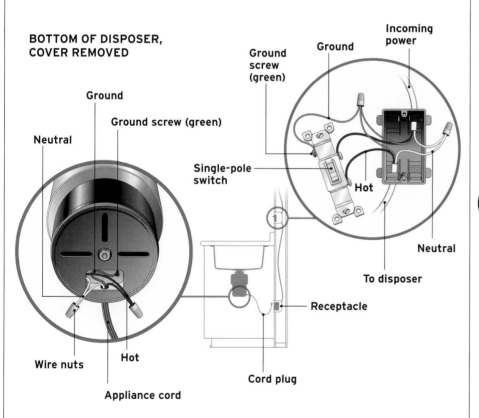

BOTTOM OF DISPOSER, COVER REMOVED

Neutral

Ground

Ground screw (green)

Single-pole switch

Wire nuts

Hot

Appliance cord

Ground screw (green)

Ground

Incoming power

Hot

Neutral

To disposer

Receptacle

Cord plug

⚠ WARNING

The NEC specifies that you connect the two hot wires of a split-tab receptacle to a double-pole breaker. When you flip off the toggle for a double-pole breaker, you shut off both hot wires. If you instead connect the hot wires of a split-tab receptacle to separate, single-pole breakers and then flip off only one breaker, you might test the top half of the split-tab receptacle and conclude—mistakenly—that the bottom half was off, too. Attaching both hot wires to a double-pole breaker prevents such a potentially lethal mistake.

ROUGHING IN AN OVEN OUTLET

Roughing in an oven outlet is not that different from roughing in any other outlet. Remove the knockout from a four-square box, insert a plastic cable connector, and then feed in the 10/3 NM cable that will power the range. Staple the cable within 12 in. of the box; drive the staple just snug **1**.

Although an experienced electrician can use a utility knife to strip sheathing from any cable, using a cable ripper that can accommodate large-gauge wire makes sense for nonprofessionals **2**. Hold the ripper channel snug to the cable and pull it down the length of the cable to score the sheathing **3**.

Pull back the sheathing to expose wires inside. Then use diagonal cutters to cut the sheathing free. Leave ½ in. to 1 in. of sheathing inside the box **4**. Attach the ground wire to the box using a green ground screw in a threaded hole. Wrap the wire clockwise around the screw so it will stay in place as the screw tightens down on it **5**.

Fold the wires neatly into the box so they can be easily pulled out during the trim-out phase, when they'll be attached to a 30-amp receptacle or hard wired directly to a metal-clad appliance whip. Finally, attach a two-gang mud ring to the outlet box so it will be flush to the finish surface **6**.

1 10/3 NM cable is run to a four-square box.

2 Use a cable ripper that can accommodate larger gauge wire.

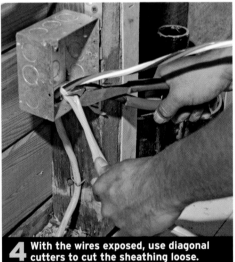

4 With the wires exposed, use diagonal cutters to cut the sheathing loose.

5 Leave the ground wire long to be spliced to the ground in the appliance whip.

ELECTRIC RANGES, OVENS, AND COOKTOPS

When discussing cooking appliances, you'll need to keep several terms straight: The enclosed cooking area in which you roast a turkey in is an *oven*; you place pots and frying pans on *cooktop* burners. A *range* has both an oven and a cooktop.

In any case, the heating elements of ranges, ovens, and cooktops generally require 240v, but today's smart appliances come with a plethora of timers, clocks, sensors, buzzers. and other gizmos that use 120v. For this reason, many units require 120/240v wiring, with two hot wires, an insulated neutral, and an equipment ground wire.

As noted elsewhere, if the unit slides in and out for maintenance, it is usually installed with a plug inserted into a matched receptacle. The outlet box that contains that receptacle may be surface mounted or recessed (so that the receptacle can be flush mounted). If the unit drops in and stays put, it is typically hard wired to a junction box via an appliance whip.

3 Pull the ripper to score the sheathing.

6 Attach a mud ring to bring the box flush.

TRADE SECRET

Appliances are usually the same depth as base cabinets (24 in.) so appliance faces will be even with cabinet faces. Because most appliances are installed against a wall, manufacturers often build a recessed area in the back of the appliance to accommodate electrical connections. Refer to the user's manual prior to rough-in.

WIRING A DROP-IN OVEN

Drop-in ovens and other stationery appliances must be hard wired. For the project shown here, an electric oven has been installed in a base cabinet, and the wires in its MC cable are ready to be connected to 10AWG wires, roughed in to a four-square box. The edge of the box must be flush to the cabinet back, because plywood is flammable. If the box or its mud ring is below the cabinet back, add a four-square box extension to bring it flush **❶**.

Install a two-piece, right-angle cable connector (a flex-90) to the end of the cable whip. Slide the bottom of the connector onto the end of the whip **❷** and screw on the top of the cable connector. Remove the stamped knockout on the four-square cover; then feed the whip wires and the connector end through the knockout **❸**. Tighten a locknut onto the threaded connector end to lock the cable connector to the cover.

To ground the outlet box, loop the incoming ground wire under a green ground screw **❹**. Splice that ground wire to the appliance ground. Use wire strippers to remove ½ in. of insulation from the neutral and hot wires on the incoming cable **❺**; then use wire connectors to splice like-colored wires together: black to black, red to red, and white to white. To ensure that wire connectors grip the wires securely, use needle-nose pliers. (There are two hot wires—red and black—because the oven requires 240v.)

Once all the wire groups are spliced using connectors **❻**, tuck them into the box. Then attach the cover to protect the connections within **❼**. >> >> >>

1 A four-square box extension brings the outlet flush.

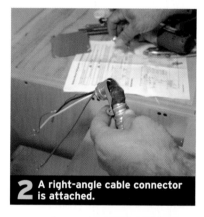

2 A right-angle cable connector is attached.

3 Remove a knockout and feed the wires through.

4 Wrap the incoming cable ground under a ground screw.

WIRING A DROP-IN OVEN (CONTINUED)

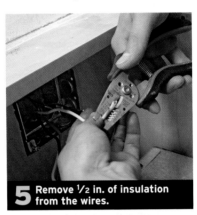

5 Remove ½ in. of insulation from the wires.

6 Fold the wires into the box.

7 Attach the cover to the box.

WIRING LAUNDRY SETUPS

Electricians sometimes run an insulated ground wire in conduit. In the setup shown here, the metal conduit is the system ground. There is, however, a green grounding pigtail that runs from each box grounding screw to a receptacle ground screw. In this project, wires are not attached to a power source, so they are safe to handle.

> → **For more on installing metal conduit, see pp. 203 and 204.**

Start by fishing wire to the box nearest to the power source. Here, four wires had to be fished: two 10AWG wires to feed the dryer and two 12AWG wires to feed the washer. Untape the wire bundle and then trim each set of wires to rough length so that at least 6 in. of wire sticks out of the box ❶; many electricians leave 8 in. to 10 in. of wire sticking out.

Attach a green grounding pigtail to the box and strip ½ in. of insulation off the ends of the incoming wires. If the point of the grounding screw hits the masonry wall, you may need to shorten the grounding screw so that it doesn't run into the masonry. Insert the stripped wire ends into screw terminals on the dryer receptacle ❷. Twist the stranded wires beforehand so they will not spread excessively when you tighten the screws. After tightening the screws, gently tug each wire to be sure it's well attached. The green pigtail grounds the device to the box.

The receptacle for the washer is a standard 20-amp receptacle. Use a wire stripper to create loops that you can attach to the screw terminals on the receptacle ❸. Once both receptacles are wired, fold the wires into their respective boxes and attach the cover plates ❹, ❺. If you install an industrial raised cover, secure the receptacle to the cover, then screw the cover to the box. Once you've installed the covers, attach incoming the wires to breakers in the panel.

>> >> >>

MEASURING PIPE LENGTH

To determine the length of conduit pipe running between two outlet boxes: Measure from the centerline each box (A).

Subtract 2 in. from the centerline of each box (B).

Subtract the distance each adaptor sticks out of the box (C).

Add the distance that pipe ends fit into adaptor sockets (D).

Pipe length = A – 2B – 2C + 2D

Distance between centerlines of boxes

A

B (2 in.)

D   D

B

C   C

Pipe end fits into adaptor socket

Adaptor (pipe connector)

Outline of four-square box

1 Trim the wires running to each box so that 6 in. to 8 in. of wire sticks out of the box.

WIRING REQUIREMENTS FOR LAUNDRY SETUPS

Clothes washers and dryers are often wired with two surface-mounted receptacles.

As with electric ranges, electric dryers typically require 120/240v wiring because, in addition to their 240v heating elements, dryers are also equipped with several elements that use 120v–such as drum motors, timers, and buzzers. So dryer circuits also include two hot wires, an insulated ground, and an equipment ground wire.

Equipment grounds on washers and dryers connect to appliance housings to provide a safe route for fault current, should a short circuit occur. Washer circuits are usually wired with 12AWG wire and protected by a 20-amp breaker or fuse; dryer circuits are wired with 10AWG wire are protected by a 30-amp breaker or fuse. But, as always, note the name-plate ratings on your appliances and wire them accordingly.

The type of appliances used deter-mines how many wires you fish. Some dryers require three incoming wires (two hot, one neutral); whereas others require two hot wires only. For the washer and dryer installed in the projects shown here, we ran four wires: two #10 stranded THHN wires (two hots) for the dryer (240v) and two #12 stranded THHN wires (a hot and a neutral) for the washing machine (120v).

2 Twist the stranded wire ends, then insert them into screw terminals.

3 Use the hole on the handle of the wire stripper to create loops.

DEDICATED VS. DESIGNATED CIRCUITS

Electrical codes require that certain critical-use appliances, such as refrigerators and freezers, be installed on a *dedicated circuit*–that is, the circuit serves only one appliance. Less well known are *designated circuits*, such as the dryer receptacle shown in the photo. The receptacle is the only one on the circuit: The distinction is that the circuit is designated for one use (laundry) but not for one appliance. Thus, if you decided to switch to a gas clothes dryer, you plug both that dryer and a clothes washer into the receptacle.

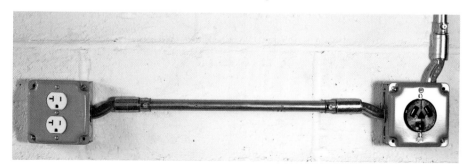

Typical laundry setup. A 30-amp dryer receptacle on a dedicated circuit (right) and a 20-amp washer receptacle on a designated circuit (left).

WIRING LAUNDRY SETUPS (CONTINUED)

4 Tuck the wires of each receptacle into the box.

5 Install the cover plates.

INSTALLING BASEBOARD HEATERS

Baseboard heaters are increasingly popular as a backup to a central heating system. Installing units with in-heater thermostats is a better choice than installing a central wall thermostat that controls all units. Baseboard units with in-heater thermostats are easier to install and more cost-effective to operate because they deliver heat to areas where it's needed most. Because most baseboard units are installed under windows, units with in-heater thermostats can respond faster to cold air as it enters.

Baseboard heaters are available in 120v and 240v models, but 240v models are generally more efficient. As a rule of thumb, you can connect several small heaters to one cable running from the main panel, as long as their combined continuous load doesn't exceed 80 percent of the cable's rating. In other words, if you run 12AWG cable with 20-amp protection, the continuous current should not exceed 16 amps. But because wattage varies from model to model, follow the installation instructions that come with your model.

Wiring an in-heater thermostat is pretty straightforward. Rough-in wiring to each heater location, remove a cover plate on one end of each unit, feed the incoming cable through a cable connector and splice the incoming circuit wires to the thermostat wires. Make sure to attach a grounding pigtail to the metal housing of the unit. Thermostats will be single-pole or double-pole switches. Double-pole thermostats are preferable because when you flip them off, you cut power to both legs of the unit, which is safer and more convenient.

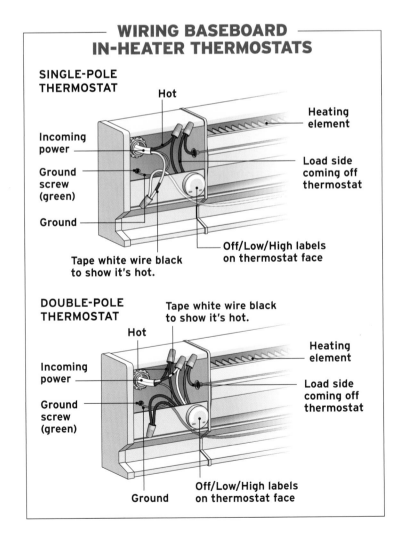

WIRING BASEBOARD IN-HEATER THERMOSTATS

SINGLE-POLE THERMOSTAT

Hot
Incoming power
Ground screw (green)
Ground
Heating element
Load side coming off thermostat
Off/Low/High labels on thermostat face
Tape white wire black to show it's hot.

DOUBLE-POLE THERMOSTAT

Tape white wire black to show it's hot.
Hot
Incoming power
Ground screw (green)
Ground
Heating element
Load side coming off thermostat
Off/Low/High labels on thermostat face

ELECTRIC WATER HEATERS

Replacing an electric water heater generally requires a permit, even if you attach the replacement to existing pipes. The reason is safety: Inspectors want to ensure that gas- and oil-fired units are properly vented and that electric heaters are correctly wired. Inspectors are particularly concerned that temperature and pressure relief (TPR) valves are correctly installed, because TPR valves keep water heaters from exploding in the event of a malfunction. For these reasons and for warranty issues, have a professional install your water heater. The drawing below is offered for information only.

TRADE SECRET

Electric tank heaters are becoming rare, given the greater cost-effectiveness of fuel-fired water heaters, especially tankless gas-fired water heaters.

WIRING A 240V WATER HEATER

Here's a look at a typical water heater installation. Three things to note: (1) The cutoff "switch" can be a circuit breaker or a fused switch rated for the load of the water heater, typically 30 amps. Place the cutoff switch close to the unit. (2) If you use two-wire cable to wire the water heater, tape the white wire with red tape on both ends to indicate that it is being used as a hot conductor. (3) Use flexible metallic cable (rather than rigid conduit) between the cutoff switch and the water heater for extra safety in earthquake regions.

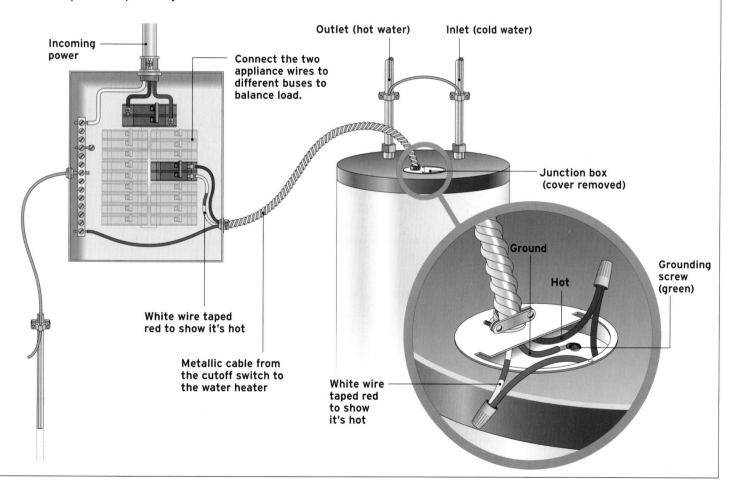

Incoming power

Connect the two appliance wires to different buses to balance load.

Outlet (hot water)

Inlet (cold water)

Junction box (cover removed)

Ground

Hot

Grounding screw (green)

White wire taped red to show it's hot

Metallic cable from the cutoff switch to the water heater

White wire taped red to show it's hot

OUTDOOR WIRING

ADDING AN OUTDOOR RECEPTACLE, a motion-sensor light, or low-voltage path lights can increase your security and safety and enable you to enjoy your property more fully. Because many outdoor devices are available as kits, they are usually easy to install and require few specialized tools. But before you start any project, review local building code requirements for outdoor installations. And remember: Before working on an existing outlet or fixture, be sure to turn off the power at the service panel or fuse box and test to be sure the power is off. Also, never operate power tools in damp or rainy conditions.

BEFORE YOU BEGIN

Planning your outdoor system, p. 224

Choosing outdoor lights, p. 225

Tools & materials, p. 226

OUTDOOR OUTLETS

Tapping into an existing outlet, p. 228

Mounting & wiring an exterior outlet, p. 230

OUTBUILDINGS

Running power to an outbuilding, p. 231

OUTDOOR PATH LIGHTS

Installing low-voltage path lights, p. 233

MOTION-SENSOR LIGHTS

Installing a motion-sensor light, p. 235

PLANNING YOUR OUTDOOR SYSTEM

Start by walking the property and noting where you'd like additional outlets, lights, and so on. Make a list. As you go, imagine activities that take place in different locations at night and day. Is there enough light along the path when you come home at night, enough outlets to entertain or to do chores on the weekend? At this stage, think big and anticipate future uses, especially if you must dig up the lawn to install your present project. With a little advance planning, you can avoid digging up the lawn a second time, later.

If your wish list is extensive, next make a scale drawing of the house and yard on graph paper. Note sidewalks, paths, and important landscape elements such as trees and large bushes. Note electrical devices that you'd like to add and existing ones that need upgrading. The drawing will be especially useful when it's time to calculate the number of fixtures and amount of cable you'll need. If you intend to take power from an existing outlet, note where the nearest outlet is, even if it's inside the house.

Getting power

Once you know roughly where you want to add exterior outlets or light fixtures, figure out how to get power to them. This will depend, in part, on how many devices you're adding.

If you're adding just one exterior receptacle, for example, find the interior receptacle closest to the one you want to add outside. If you position the new receptacle in the same stud bay as the interior receptacle, you'll simplify the task immensely because you won't have to drill through any studs.

Before cutting holes in anything, however, calculate the load on the circuit. Add up the wattage of all the lights and appliances presently in use and the wattage of the new outlet or light you want to add. If the total load on a 15-amp circuit exceeds 1,440w, run a new grounded circuit from the panel to the exterior device instead. Likewise, if you are running power to an outbuilding, run a new grounded circuit from the panel.

Install a 20-amp circuit if it will contain four or more receptacles or three or more large flood lamps or other large lighting fixtures rated more than 300w each. If you will be using large, stationary tools in a workshop or heating the area with electricity, you may need to add several 120v and 240v circuits. Again, calculate the loads involved.

Two important points: First, the NEC allows you to tap into a general-use or lighting circuit only. You may not tap into circuits feeding kitchen-countertop receptacles or bathroom receptacles or into any dedicated circuits that supply power to air-conditioners, clothes washers or dryers, and so on. Second, if you discover that the circuit you'd like to tap into is wired with nongrounded NM cable, BX armored cable, or knob-and-tube—don't tap into that circuit. The new extension may not be properly protected. Instead, run a new grounded circuit from the panel.

Finally, take your plans for new circuits or extensions to existing line-voltage circuits to the local building department and have them sign off before you start.

Outdoor safety: checking codes and utilities

Outdoor outlets and fixtures are exposed to weather; because moisture greatly increases the chance of electrical shocks, local codes are strict about what materials you can use and how they must be installed. In general, you don't need a permit to install low-voltage lights because the chance of shock is low; but if you want to add an outdoor receptacle or a light fixture that uses line voltage (standard house current of 110v or 220v), you'll need a permit.

If you'll be running cable underground, check with local utilities before you dig. Call USA (Underground Service Alert) 800-227-2600. There may be water pipes, gas lines, telephone or cable lines, or electrical cables buried in the yard. Often, utilities will send out a technician to show you where such lines are located. If your lawn has a sprinkler system, note where sprinkler heads are and try to avoid the water pipes that feed them. Remember that only a licensed electrician should install hot tubs, swimming pools, and the like because such installations require special grounding methods.

CHOOSING OUTDOOR LIGHTS

There are a wide variety of light fixtures to choose from. For starters, choose line-voltage lights (120v) if you want to deter intruders, accent an architectural feature, or illuminate a work area such as an outdoor grill. To light up a walkway or add accent lights to the landscaping, however, low-voltage lights (12v or 24v) are usually a more economical choice and are generally easier to install. There are also lo-vo solar units (no wiring needed) that charge during day and glow softly all night.

In general, don't install more light than you need to serve the function for a given area. Outdoor lights that are too bright waste energy and will be too glaring for intimate dining or entertaining at night. Your neighbors will also thank you for not spot-lighting their house when they're trying to relax or sleep.

In addition to overhead lights, side-mounted lights, step-riser lights, in-ground fixtures, post-mounted lights, and stake-mounted lights, there are many switching options. You can control lights with standard on-off switches, timers, motion detectors, and photocells that turn lights on when the sun goes down. Most home centers and lighting stores have a dazzling array to choose from.

Outdoor lights are available in an array of styles. Here, Mission-style lights brighten a stone stairway.

ILLUMINATING OUTDOOR LIGHTS

As for your own safety, remember that intruders dislike being in the limelight. Install a security light, and they'll probably go elsewhere. A few tips to make your lights more effective:

- Install motion-sensor lights. Because they are dark until triggered by motion, motion-sensor lights startle intruders. Better-quality sensors can be calibrated so they are activated only when someone nears the house—not by every dog walker on the block.

- Put security lights high on a porch or under the eaves. Lights that can be reached without a ladder can be easily unscrewed. Most security lights have dual sockets so that if one bulb burns out, there's still one shining.

- Light walks and doors. It's easier for you to quickly approach and enter your house if sidewalks and entry doors are lit. Groping for your house keys in the dark is especially ill-advised if you live alone or along a dimly lit street.

- There are also security light kits that turn on lights inside the house if someone approaches or touches a window. And if you're leaving the house for an extended period, put lamps on timers to confuse would-be intruders.

Motion-sensor lights not only welcome you home but also help keep intruders away.

TOOLS & MATERIALS

A Use a square-nosed shovel to create shallow slots so you can tuck lo-vo cable into the ground.

B Rent a ditch-digging machine, also known as a trencher, to save time and avoid a sore back.

The tools you need to install outdoor wiring are pretty much the same ones needed to wire a house interior. The big exceptions, obviously, are digging and earth-moving tools. A *square-nose shovel* is the most useful tool when you're making a shallow slot for low-voltage cable **A**. If you're actually digging a trench, use a trenching shovel with a pointed nose and a reinforced shoulder that you can stamp on with a boot to drive it deep.

Digging trenches is hard work, however; so when the pros have to dig one of any length they rent a gas-powered ditch-digging machine, also known as a *trencher* **B**. A trencher looks like a cross between a powered garden tiller and a chainsaw and typi-

cally cuts a trench 4 in. to 6 in. wide, and as deep as 24 in. Spread sheet plastic on either side of the trench so you can place the dirt from the trench nearby—which makes refilling the trench easier. Wear heavy boots, heavy gloves, safety glasses, and ear plugs when operating a trencher.

Materials

The NEC requires that all exterior outlets and circuits have ground-fault interrupter (GFCI) protection. This protection includes the following:

■ A GFCI receptacle installed in a weatherproof box
■ A standard receptacle installed in a weatherproof box protected by a GFCI

receptacle installed *upstream* (toward the power source) or by a GFCI circuit breaker
■ A circuit protected by a GFCI breaker.

Exterior light fixtures do not need GFCI protection. If the box is surface mounted, it needs to be raintight. If the box is recessed in the wall, it does not. The fixture must be listed for *damp locations* if under eaves and listed for *wet locations* if directly exposed to weather.

There are two common types of weather-proof covers (also called raintight covers) (see top photo on the facing page). A *weatherproof-while-in-use cover* has a plastic cover that shuts over an electrical cord, such as that used for holiday lights. A *weathertight cover* is a gasketed cover that shuts tight over the receptacles when not in use.

Aboveground, electrical cable must be housed in conduit with raintight fittings (see bottom photo at right). You can use polyvinyl chloride (PVC), thin-walled electrical metallic tubing (EMT) conduit, or threaded intermediate metal conduit (IMC) or rigid steel conduit (RSC). PVC fittings are glued together to achieve a raintight fit; EMT conduit uses compression fittings; and IMC and RSC use threaded fittings. *Note:* EMT conduit fittings intended for interior use are not raintight and are *not* approved for exterior use.

Belowground, you can run flexible underground feeder (UF) cable at a depth acceptable to local codes—typically, 18 in. to 24 in. deep. Some codes allow you to dig a shallower trench if the cable runs in steel conduit but because threading steel conduit requires special equipment and advanced skills, it's not a reasonable option for most nonprofessionals.

WARNING

Avoid plugging or unplugging devices into exterior receptacles—or using corded power tools outside—when it's raining or snowing.

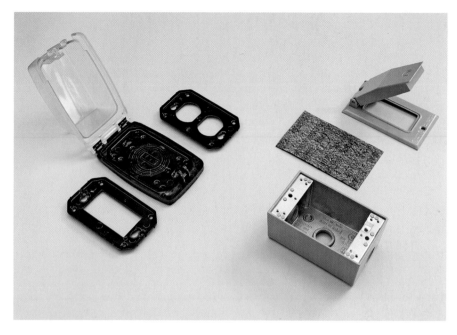

Raintight (weatherproof) covers include a weatherproof-in-use receptacle cover with adapter plates (left) and a weatherproof box with gasketed cover (right). Plastic adapter plates enable you to use the cover with a duplex or GFCI receptacle. The weatherproof box gasket is precut, so it accepts either a duplex or GFCI receptacle.

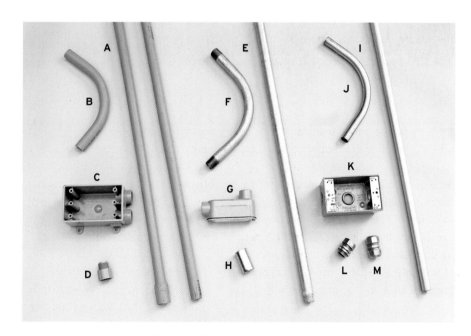

Overview of exterior conduit types, fittings, and boxes: A, 1/2-in. PVC pipe; B, PVC elbow; C, PVC box with unthreaded openings (for slip-in fittings); D, male adapter (MA) PVC fitting, which can be used with a threaded box or conduit; E, RSC pipe; F, RSC elbow; G, RSC coupling; H, LB conduit; I, EMT pipe; J, EMT elbow; K, bell box with three threaded holes; L, threaded EMT compression fitting; M, EMT compression coupling. Note: The LB conduit and bell box accept any 1/2-in. (trade size) threaded fitting—whether PVC adapter, EMT, or RSC.

TAPPING INTO AN EXISTING OUTLET

1 Remove the cover plate and unscrew the receptacle from the outlet box.

2 Pull the receptacle out of the box and remove the wires attached to its terminal.

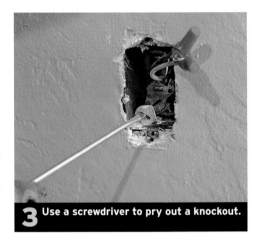

3 Use a screwdriver to pry out a knockout.

7 Pull the new wire through and attach a cable clamp.

8 Leaving one leg long, use a green grounding nut.

When adding a single outdoor receptacle, it's usually easiest to tap into an interior outlet within the same stud bay. In the installation shown here, the electrician solved the box-capacity problem by replacing the old receptacle with a commercial-grade duplex receptacle. Instead of splicing new and old wires and using pigtails—which would have required twist-on wire connectors and thus a larger box—he joined incoming and outgoing wires by inserting them into terminal holes in the back of the receptacle and then tightened down the terminal screws. *Note*: Wiring a commercial-grade receptacle in this manner is not the same as back-wiring a standard receptacle because standard receptacles have inferior spring clamps that are unreliable.

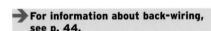

➡ **For information about back-wiring, see p. 44.**

Start by turning off the power to the receptacle at the panel or fuse box; then use an inductance tester to make sure the power is off. Remove the cover plate and unscrew the mounting screws securing the receptacle to the outlet box **1**.

Pull the receptacle out of the box, being careful not to touch the screw terminals; then test them again to make sure they're not energized **2**.

Detach the wires from the terminals, fold them out of the way, and look for a knockout in the box to remove. Typically, a pair of knockouts will be hidden beneath an integral cable connector—unscrew it. Then remove a knockout from the bottom of the outlet box **3**. Drill through the exterior wall and into the same stud bay as the interior receptacle **4**. Then, inside the house, push fish tape through the knockout hole and into the stud bay **5**. Outside, a helper can catch the tape, pull it out of the hole you just drilled, and attach the new cable to it **6**.

Inside, pull the fish tape and the attached cable into the box. Pull about 1 ft. of new cable out of the box and then replace the integral cable clamp **7** to secure the cable. Strip the cable sheathing and splice like wires together, starting with the ground

wires. Use a special green grounding nut to splice the grounds **8**. Use wire strippers to remove 1/2 in. of the insulation from the ends of the hot and neutral wires **9**.

To save space in the old box, insert the stripped wire ends into terminal holes in the back of the commercial-grade receptacle **10**. Because commercial-grade receptacles solidly clamp wire ends, this connection is as solid as any splice. Fold the wires into the box as you push the receptacle into place until it is fully seated in the box—do not use mounting screws to pull the receptacle into the box because this could strip the screw threads. Then replace the cover and screw it into place.

4 Outside, have a helper drill through the siding.

5 Push the fish tape through the knockout hole, into the stud bay.

6 Outside, retrieve the fish tape and attach the new wire.

9 Use wire strippers to remove 1/2 in. of the insulation.

10 Gently push the wired receptacle into the box.

WARNING

To power holiday lights safely, plug them into a GFCI receptacle housed in a water-proof-while-in-use box cover. But before installing the lights, calculate the total wattage of all the bulbs so you don't overload the circuit.

ADDING AN OUTDOOR RECEPTACLE

If you tap into an existing receptacle at the end of a circuit, there should be enough room inside the box to bring a cable to feed the new outdoor receptacle. However, if there are already two cables in the box—incoming and outgoing—you may need to replace the existing box with a larger one.

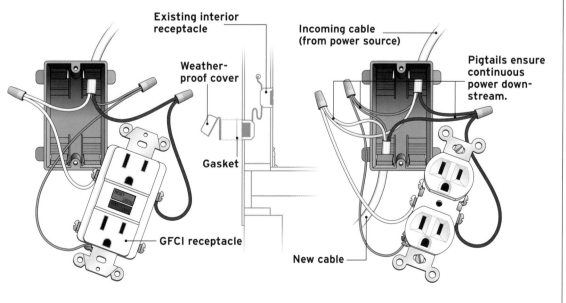

Existing interior receptacle

Incoming cable (from power source)

Pigtails ensure continuous power downstream.

Weather-proof cover

Gasket

GFCI receptacle

New cable

MOUNTING & WIRING AN EXTERIOR OUTLET

1 Feed the cable through the connector in the box and secure it to the building.

2 Loop the bare ground wire around the green grounding screw.

3 Attach the long ground wire to the green ground screw.

4 Insert wires into screw terminals. Gold screw for hot wire; silver screw, for neutral wire.

5 Fold and tuck the wires into the box. Then secure the receptacle to the box.

6 Fit the gasket around the receptacle and screw on the weatherproof cover.

Once the new cable is spliced to the cable in an existing receptacle, feed the cable through the cable connector in the back of the exterior box and mount the box. (The hole drilled in the exterior wall must be wide enough for the cable connector or the box won't sit flat to the wall.) To mount the box, use mounting ears or the small holes in the back. Apply siliconized caulking to the hole before attaching the box. Plumb the box and screw it to the outside of the building **1**.

Strip sheathing from the cable and ground the box by looping the bare ground wire around the green grounding screw. Leave this ground wire long **2**. Use a wire stripper to remove ½ in. of the insulation from the ends of the neutral and hot wires. Attach the ground wire that you earlier looped around the green ground screw to the GFCI receptacle **3**. Then connect the hot and neutral wires in their respective screw-terminal holes **4**.

If the GFCI receptacle has plaster ears, use diagonal cutters to remove them; otherwise, the receptacle may not fit into the box. Fold the wires and push the receptacle into the box; then screw down the mounting screws that secure the receptacle to the box **5**. A weatherproof gasket is used to keep water away from the wires—set it in place around the receptacle before you attach the cover **6**.

TRADE SECRET

GFCI receptacles are larger than standard duplex receptacles, so there will not be enough room in a single-gang box if you also need to splice an outgoing cable to feed another outdoor outlet, downstream. In that case, install a deep box or install an extension to the single-gang box.

RUNNING POWER TO AN OUTBUILDING

The first step to wiring an outbuilding is to figure out how many lights and outlets you need. If your needs are modest, you may be able to tap into an existing outlet in the main house and extend the circuit from it. Calculate the total loads for the existing circuit and the extension to see if the circuit has enough capacity. Otherwise, run a new circuit from the panel to the outbuilding.

If the outbuilding isn't more than 50 ft. from the house and has a few lights and outlets, it usually can be supplied by a 120v, 20-amp circuit and 12/2 w/grd UF cable. But check with local building authorities before you start. Get the necessary permit, Code requirements, and inspection schedules.

The hardest part of the job is usually digging the trench because local codes typically require it to be 18 in. to 24 in. deep. Fortunately, you can rent a gas-powered trencher to do the digging for you. After removing the dirt from the trench, pick out any rocks or debris that could damage the cable. Then lay the UF cable in the trench, flattening it as you go ❶.

>> >> >>

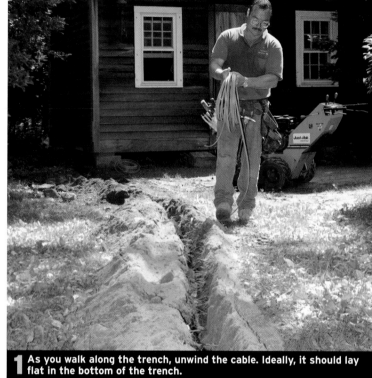

1 As you walk along the trench, unwind the cable. Ideally, it should lay flat in the bottom of the trench.

RUNNING POWER TO AN OUTBUILDING

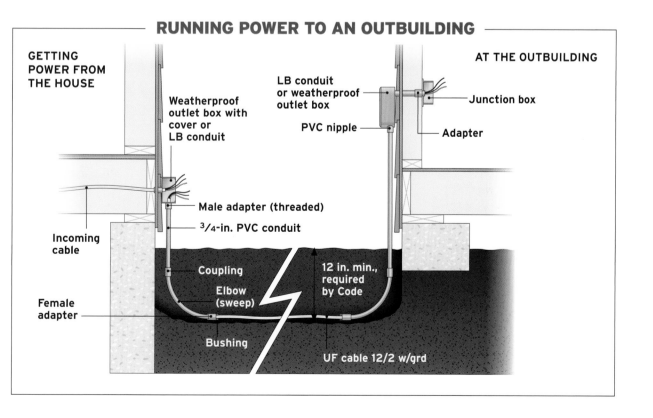

GETTING POWER FROM THE HOUSE

AT THE OUTBUILDING

Weatherproof outlet box with cover or LB conduit

LB conduit or weatherproof outlet box

Junction box

PVC nipple

Adapter

Male adapter (threaded)

³/₄-in. PVC conduit

Incoming cable

Coupling

12 in. min., required by Code

Elbow (sweep)

Female adapter

Bushing

UF cable 12/2 w/grd

RUNNING POWER TO AN OUTBUILDING (CONTINUED)

Use electrical-grade PVC conduit, couplings, and adapters to protect the UF cable between the bottom of the trench and the outdoor boxes in which connections will be made. At the house, UF cable is typically spliced to an interior wire in a covered single-gang, weathertight box ❷. On the other end, the cable typically passes through an LB condulet and a short length of PVC pipe before terminating in a junction box inside the outbuilding ❸.

2 A PVC stub protects UF cable as it emerges from the ground.

3 Strap rigid PVC pipe (with LB conduit) to the outbuilding.

WORKING WITH PVC CONDUIT

PVC conduit can be heated and bent, allowing you to change pipe direction with far less cutting and fewer fittings than you would need for rigid pipe. PVC conduit is intended to be used outside: It is flexible, durable, and waterproof; but its assemblies are not good-looking or as protective as metal pipe or cable. Be sure to use only the gray PVC conduit rated for electrical work (also called schedule-40 PVC conduit). White PVC plumbing pipe is unsuitable as electrical conduit; moreover, heating it can release noxious fumes. Wear work gloves to avoid burns when heating gray PVC pipe.

It's best to use a hacksaw with a metal-cutting blade to cut PVC pipe, though any fine-tooth saw will work. Make the cut as square as possible. Use a pocket knife or curved file to remove burrs from the pipe end, then wipe it with a rag. Apply an even coat of PVC cement to the outside of the pipe and the inside of the fitting. If you're bending pipe, screw the threaded adaptor into the outdoor box or conduit first, then glue the bent pipe to the adapter.

To bend PVC pipe, heat it with a MAP-gas torch; one common brand is the Bernz-Omatic™. You can also use a hot box or a PVC heating blanket (Greenlee). Keep the torch point moving constantly, rotate the pipe periodically, and tape the pipe ends to contain the heat and speed up the process. When the pipe droops, it's ready to shape. Place it against an irregular wall or foundation and it will conform to that contour. Strap the pipe close to the outdoor box and the foundation. To secure straps to the foundation, first predrill with a masonry bit. Then insert expandable plastic anchors into the holes and drive strap screws into the anchors.

To bend PVC pipe, heat it with a handheld MAP-gas torch. Rotate the pipe and keep the torch moving to heat the pipe evenly.

When the heated PVC pipe begins to droop, you know it's ready to bend. Shape it to fit the side of the building or foundation.

Use straps to secure the pipe to exterior walls and foundations.

INSTALLING LOW-VOLTAGE PATH LIGHTS

Low-voltage lights make nighttime paths and walkways safer, are easy to install, and pose almost no shock threat. Always follow the installation instructions that come with your low-voltage kit. Kits usually include light assembles, lo-vo cable, posts, ground stakes, and a timer-transformer power pack whose transformer reduces house current from 120v to 12v.

In the installation shown here, the power pack was mounted inside the garage, so PVC conduit was installed to protect the lo-vo cable as it traveled up the exterior wall into the garage. If you install the power pack outside, you probably won't need conduit. If you do install conduit, start by inserting a fish tape down the conduit ❶. Separate the two wires in the lo-vo cable, snip one, loop and tape the remaining wire to the fish tape, and pull the lo-vo cable through the conduit.

Place the lights where they'll best illuminate a walkway or highlight a landscape fea-

ture, then run the lo-vo cable to them ❷. Run cable along the ground and cover it with a few inches of mulch, or use a square-nosed shovel to create a shallow slot for the cable. Stomp on the shovel so it goes down 4 in., then rock the shovel from side to side to create a V-shaped slot ❸. Press the cable into the slot ❹; then stamp your feet to close the soil over the cable.

Each light fixture has two wire leads that terminate in sharp-pointed cable connec-

>> >> >>

Lo-vo light kits typically contain **screw-together parts: lamps, shades, riser posts, and ground stakes. At right: a coil of lo-vo cable and a power pack—a combination timer and transformer to step down voltage to 12V.**

1 After installing PVC conduit, insert a fish tape down the conduit and pull the wires through.

2 Place the lights and run the conduit to each of them.

3 Use a square-nosed shovel to create a small slot to bury the cable.

4 Press the lo-vo cable into the slot, making sure that it's fully covered.

INSTALLING LOW-VOLTAGE PATH LIGHTS (CONTINUED)

tors. When snapped together, the connectors pierce the insulation of the lo-vo wires and supply each light with 12v current **5**. To power the system, attach lo-vo cable wires to the terminals on the power pack **6**. Mount the power pack securely to the wall and plug it into the outlet. Power packs are equipped with timers so that the lights come on and turn off whenever you like—whether you're home or away **7**.

5 Connectors are used to join the cable in the ground to each light.

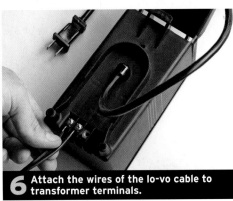

6 Attach the wires of the lo-vo cable to transformer terminals.

7 Once hung, plug the power pack into a nearby outlet and set the timer.

LOW-VOLTAGE LIGHT PARTS

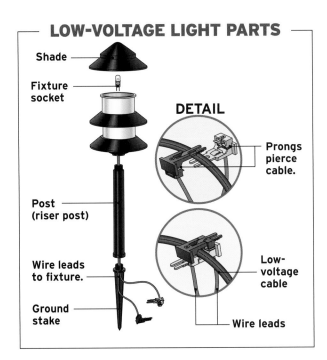

Shade

Fixture socket

DETAIL

Prongs pierce cable.

Post (riser post)

Wire leads to fixture.

Low-voltage cable

Wire leads

Ground stake

RUNNING LO-VO CABLE UNDER A SIDEWALK

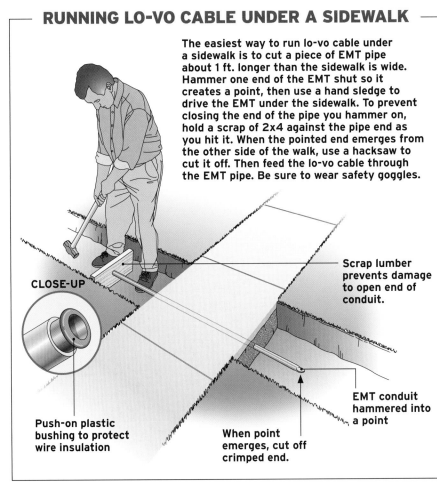

The easiest way to run lo-vo cable under a sidewalk is to cut a piece of EMT pipe about 1 ft. longer than the sidewalk is wide. Hammer one end of the EMT shut so it creates a point, then use a hand sledge to drive the EMT under the sidewalk. To prevent closing the end of the pipe you hammer on, hold a scrap of 2x4 against the pipe end as you hit it. When the pointed end emerges from the other side of the walk, use a hacksaw to cut it off. Then feed the lo-vo cable through the EMT pipe. Be sure to wear safety goggles.

Scrap lumber prevents damage to open end of conduit.

CLOSE-UP

Push-on plastic bushing to protect wire insulation

When point emerges, cut off crimped end.

EMT conduit hammered into a point

INSTALLING A MOTION-SENSOR LIGHT

Motion-sensor lights require 120v, so they must be mounted on boxes rated for outdoor use. The fixture must be listed for damp locations if installed under eaves or must be listed for wet locations if directly exposed to weather. If you are replacing an existing light with a motion-sensor light, don't assume that the old box is raintight—examine it. Exterior boxes should have flexible gaskets between the box and the fixture base and threaded openings with closure plugs. If the box has only standard knockouts, it's not raintight.

Turn off power to the fixture and test to be sure it is off. Then unscrew the mounting screws securing it to the box. Holding the fixture in one hand, touch an inductance tester to the wire splices to be sure they are not energized. Disconnect the wire splices; if the existing box is damaged or inappropriate, remove it. The new box must be raintight ❶.

Feed the incoming wires through the cable connector in the back of the box and then attach the box to the wall ❷. Mounting screws should sink into sheathing or into a block screwed to the sheathing. Attach the mounting bar to the box ❸ and attach a ground wire to the ground screw on the box or mounting bar. Splice the ground wires from the box and the incoming cable ❹.

Fit a weatherproof gasket over the cable wires; then splice them to fixture leads: hot to hot, neutral to neutral ❺. Tuck the wires into the box and attach the fixture ❻. Before screwing the bulbs into the sockets, slide waterproof washers onto them ❼. Adjust the motion sensor so that it is triggered by people approaching the house—not by passing cars, dog walkers, and the like.

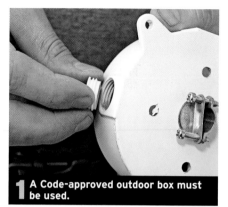

1 A Code-approved outdoor box must be used.

2 Feed incoming wires through the back of the box; then attach the box.

3 The green ground screw ensures a continuous ground path to the box.

4 Splice all ground wires.

5 Slide a weatherproof gasket over the incoming cable wires.

6 Fold all of the wires back into the box; attach the fixture.

7 Waterproof washers around each bulb prevent water from getting into the sockets.

PANELS & SUBPANELS

THIS CHAPTER OFFERS AN overview for those who want to add a circuit to an existing panel, learn about installing subpanels, or understand the major elements of a service upgrade. Adding a circuit is not difficult, but we recommend having a licensed electrician do all work inside a panel. In particular, only a professional should upgrade electrical service. If you want more information about panel work or service upgrades, hire a pro and watch at a safe distance.

There are several reasons for this caution: First, there's no obvious way for a nonprofessional to be sure that an existing system is correctly grounded and free from shock dangers. Second, some panels are so old, poorly wired, or overloaded that simply removing or replacing a cover can be dangerous. Third, local building codes may require that major upgrades be done only by licensed electricians. Last, an inexpertly wired electrical system may nullify appliance or equipment warranties or even void your homeowner's insurance should an electrical fire occur.

BEFORE YOU BEGIN

Understanding service panels & subpanels, p. 238

Sizing panels, subpanels, & conductors, p. 240

CIRCUIT BREAKERS

Adding a circuit breaker, p. 242

SUBPANELS

Installing a subpanel, p. 244

UNDERSTANDING SERVICE PANELS & SUBPANELS

Electrical panels vary in size and configuration, contributing to the general confusion over how to wire them correctly. So let's first look at how utilities supply electricity (service entrances) and then consider the difference between *service panels* (also called main panels or main service panels) and *subpanels*.

For decades, overhead service entrances were the only type, with large service conductors running from a utility pole to a *service head* (or weather head) mounted on or near the roof. But large wires are unsightly, so buried service entrances have become more and more popular.

Typically, three service conductors run to a meter base. The two hot conductors attach to lugs on the supply side of the meter, and the neutral conductor attaches to a neutral bus bonded to the metal box. When a meter is inserted in the base, power flows from the house-side meter lugs to a main disconnect (main breaker) in a service panel. Important point: All service panels contain a main breaker. Subpanels typically do not, unless they are housed in a separate building from the main breaker.

In most setups the main breaker attaches directly to two hot buses. Turn off the main breaker and you disconnect power to all breakers and circuits energized by those buses and to any subpanels *downstream*—away from the power source. Nonetheless, always test to be sure the power is off after flipping off the main breaker.

To be safe, an electrical system must be grounded. At the transformer, a ground wire runs from the neutral conductor to a ground rod driven into the earth. At the service panel, a ground wire attaches to a ground bus,

WORKING SAFELY IN A SERVICE PANEL

Cut power inside the service panel by flipping the main breaker off or by removing the main fuse in a fuse box. This deenergizes the hot buses. But remember that large feeder wires may still be hot on the incoming side of the main breaker—avoid touching that area. Unscrew the panel cover and set it aside.

Carefully test feeders and breaker terminals for voltage. If voltage on any exposed part is still present, carefully replace the cover and call a licensed electrician. If no voltage is present you can preoceed. If you have any doubt or uncertainty, call a licensed electrician. Do not take the risk!

Always turn off the power before working on an electrical system.

After removing the cover, touch one tester prong to the bare end of a feed wire and the other prong to the neutral bus.

Avoid touching the feeder wires that attach to the main breaker—they stay hot even when the main breaker is turned off.

WARNING

If you are at all uncertain whether a panel is energized, do not remove its cover. Call a licensed electrician. Do not take the risk!

LOCK 'EM OUT!

Once you've shut off power in a service panel, tape the panel shut and post a warning sign of work in progress as good first steps. But it's still possible for someone to remove the tape and reenergize the system. Pros prefer to use a *breaker lockout*, which limits panel access to the person holding the key. You can buy panel lockouts at electrical suppliers and most home centers.

exits through the bottom of the panel, and clamps to a ground rod. If there are separate neutral and ground buses in the service panel, both will be bonded to the metal body of the panel. In some service panels, both neutral and ground conductors attach to a common neutral/ground bus.

This is a second important distinction: Neutral and ground conductors are bonded at the service panel, but *never* in a subpanel. That is, in a subpanel there will always be a separate neutral bus and a ground bus. Typically, the ground bus bar in a subpanel will be bonded to the metal body of the panel, whereas the neutral bus bar in a subpanel will be mounted on nonconductive brackets.

Again, downstream from the service panel, grounds and neutrals are always separate. Thus, although three service conductors feed a service panel, four conductors feed a subpanel: two hot wires, one neutral, and one ground.

SERVICE ENTRANCE TO THE EAVES SIDE

When the service drop approaches over eaves, send the service riser up through the eaves. Apply silicone caulk where the riser emerges from the roof jack.

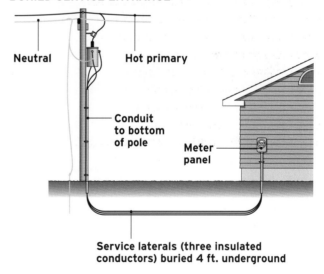

SERVICE ENTRANCES

Though most older homes have overhead service, buried service is becoming increasingly common. The utility's responsibility ends where the service-drop cables are spliced to the service conductors running to the meter. In some areas, Code requires that underground conductors be housed in conduit.

AERIAL SERVICE ENTRANCE

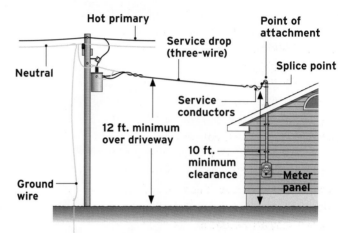

BURIED SERVICE ENTRANCE

SIZING PANELS, SUBPANELS & CONDUCTORS

A METER-MAIN COMBO PANEL

Inside a meter-main combination panel, before the meter has been installed.

Hot service conductors

Rigid conduit and service riser

Hot service conductors

Main breaker

Subpanel breaker

Meter base

Neutral/ ground bus (service panel only)

Hot buses

Neutral feed to subpanel

Hot feeds to subpanel

Neutral service conductor

Grounding wire to ground rod

Ground feed to subpanel

Protective plastic bushing

A SUBPANEL WITH FEEDS TERMINATED AT LUGS

If there's no room left in the main panel, adding a subpanel is a good option.

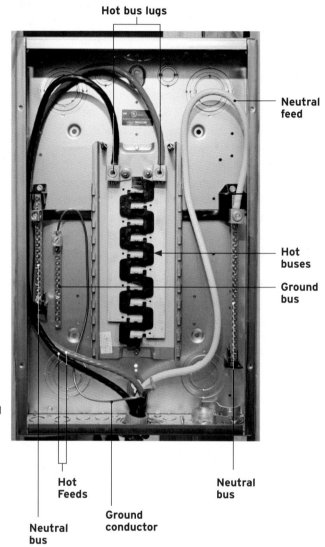

Hot bus lugs

Neutral feed

Hot buses

Ground bus

Neutral bus

Hot Feeds

Ground conductor

Neutral bus

To size the service panel, many electricians calculate household needs (ampacity), then add 20 percent to 25 percent for future needs.

➤ For more on calculating household needs, see p. 164.

Because lifestyles and energy consumption vary widely, it's tough to generalize about service panel size. For a family in a 2,000-sq.-ft. house, a 125-amp service panel is probably adequate. But many electricians recommend a 200-amp panel for an "average" household in a home less than 3,500 sq. ft. that is not heated electrically.

The NEC allows a maximum of 42 breaker spaces in a single panel. If your system still has capacity but you've used up the available panel spaces, add a subpanel. If your needs exceed system ampacity, upgrade the service.

All conductors—whether they feed a service panel or energize wall outlets—must be sized according to the loads they carry. If you'd like more information on sizing conductors you can either get a copy of the most recent *National Electrical Code* or a copy of *Code Check: Electrical* (Taunton Press), which does a good job of summarizing the NEC tables.

To give one example of service conductor sizing: A home with a 200-amp service panel should be powered by three 2/0 copper THHN/THWN conductors. As noted earlier, two of the conductors are hot and the third is neutral (taped white). Each hot conductor carries 120v relative to the neutral, so together the two hot conductors are capable of delivering 240v relative to each other. Remember, only licensed electricians and utility company workers should upgrade a service entrance.

Key Terms

Conductors
Technically, anything that conducts electricity is a conductor, but the term most often denotes individual wires, regardless of size. Thus the large wires that run from a utility pole are service conductors. The fat wires that run down a service riser to a meter base are conductors, as are the slim circuit wires that energize outlets and light fixtures.

Wire
Wire is a generic term that refers to an individual conductor, and it is most often used when referring to the size or type of the conductor, for example 12AWG and 2/O THHN/THWN stranded wires.

Cable
Cable is an assembly of several conductors, usually in a plastic or metal sheathing. Hence, Romex cable, metal-clad (MC) cable, SE cable, and so on. SE cable, a frequently misused term, is a specific type of large, non-metallic cable that is *not* an acceptable service conductor in all regions.

Feeds
Feeds, or feeders, are distinguished by function more than form, though typically they are larger conductors that supply power to an element downstream. Thus feeders might run from a meter base to a main breaker or from a service panel to a subpanel.

ADDING A CIRCUIT BREAKER

Note that 20-amp twin breakers **are often installed to save panel space.**

1 Remove a knockout from the existing panel.

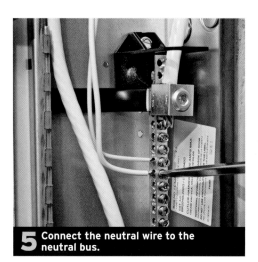

5 Connect the neutral wire to the neutral bus.

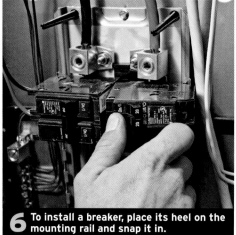

6 To install a breaker, place its heel on the mounting rail and snap it in.

7 Attach the hot wire to the breaker screw terminal.

Circuit breakers come in various configurations. In the project shown here, a 20-amp twin breaker is being installed to conserve space. (A *twin* is two slim single-pole breakers in a standard single-pole breaker space.) A typical single-pole breaker is installed in exactly the same way.

When adding a circuit to an existing panel, cut the power to the panel and test to be sure it's off. Remove the cover, then use needle-nose pliers to jab out a knockout **1**. Install a cable connector into the knockout, pull the cable through the connector, and tighten the connector to prevent strain on electrical connections **2**. Cut enough extra cable to reach the farthest point inside the panel. Staple cable within 12 in. of the panel.

Score the cable sheathing by making two diagonal passes and sliding the sheathing off. Or score the sheathing lightly down the middle **3**, peel back the sheathing, and use diagonal cutters to snip it off. Leave about 1/2 in. of sheathing sticking into the panel.

Separate the ground, neutral, and hot wires. Connect the ground wire first **4**. In a service panel, the ground will attach to a ground/neutral bus; in a subpanel, the ground will attach to a separate ground bar. Insert only one ground wire beneath each lug screw. Next strip 1/2 in. of insulation from the neutral and connect it to the neutral bus **5**. Making right-angle turns in individual neutrals toward the bus creates a neat, orderly arrays of wires.

Snap the breaker onto the hot bus **6** and press down to seat the breaker securely on the hot bus pole(s). Strip 1/2 in. of insulation from the hot wire, then connect it to the lug screw on the end of the breaker **7**. *Note:* Before attaching the hot wire, slide a circuit tag onto the hot wire to identify the location it feeds.

After connections in the panel are complete, make sure all splices, switches, receptacles, fixtures, and/or appliances on the new circuit are installed and there are no open wires or shorts in the circuit. Replace the cover, turn on the breaker, and test for proper function. Replace the cover, and turn on the power.

2 Insert the threaded shaft of the cable connector (clamp) into the knockout.

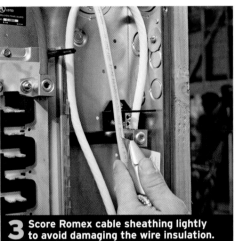

3 Score Romex cable sheathing lightly to avoid damaging the wire insulation.

4 Connect the ground wire to the ground bus of the subpanel.

GFCI VS. AFCI BREAKERS: WHAT'S THE DIFFERENCE?

Installing GFCI and AFCI breakers is essentially the same, but these two Code-required breakers offer protection from different hazards.

Ground fault circuit interrupter (GFCI) breakers appeared first. A ground fault is any failure of the electrical system that leaks current from a hot wire. GFCIs are highly sensitive devices that can detect minuscule (5 milliamp) current leaks and shut off power almost instantaneously—typically, within 1/40 second. A GFCI's primary function is to protect people from electrical shocks, especially when there is moisture present. The NEC requires GFCI protection on all bathroom receptacles; kitchen receptacles within 4 ft. of a sink; all receptacles serving kitchen counters; outdoor receptacles; accessible basement or garage receptacles; and receptacles near pools, hot tubs, and so on.

Arc-fault circuit interrupter (AFCI) breakers, on the other hand, help prevent house fires. AFCIs detect minute fluctuations in current associated with arcing and deenergize the circuits before a fire can start. Arcing most commonly occurs when a nail or screw punctures a cable, when insulation on an extension cord becomes frayed, or connections become loose at an outlet or a switch. The NEC requires AFCI protection on all 120v bedroom circuits.

DOUBLE-POLE CIRCUIT BREAKERS

The back of a double-pole breaker has two pressure clips that snap onto two hot bus poles—thus providing 240v of power. With double-pole breakers, both red and black hot wires are attached to the breaker. Energy will flow whether you connect a red or black hot wire to either terminal, but electricians routinely alternate colored wires: black, red, black, red.

INSTALLING A SUBPANEL

There are many reasons to install a subpanel:

- Increasingly, main service panels are installed outside so firefighters can disconnect power before going into the house—in this case, a subpanel inside contains all the branch circuits.
- When a system has unused capacity but the service panel has no available slots for more circuit breakers, a subpanel allows for expansion.
- A subpanel distributes power to a separate building.

- A subpanel can offset voltage drops on circuits that are too distant (70 ft. or 80 ft.) from the service panel. In this case, the larger gauge wires that feed subpanels suffer less voltage drop than the smaller gauge wires of branch circuits.

Size subpanels based on anticipated loads. If you're adding a subpanel in the same building as the main panel so you can add lighting and general-use circuits, install a 60-amp subpanel with at least 12 breaker slots. If the subpanel is distributing power to a distant building that you want to use as a workshop or office, or perhaps expand later, install a 100-amp or 150-amp subpanel.

People frequently add subpanels when planning a major kitchen remodel, because kitchens have a lot of appliances. Locate the subpanel as close as possible to the kitchen.

When adding a subpanel to an existing system, an electrician first shuts off power at the service panel and tests to be sure the power is off.

If studs are spaced 16 in. on-center, install a standard, 14$\frac{1}{2}$-in.-wide panel. Use four 1$\frac{1}{4}$ in. by 10 screws to mount the panel ❶; screws should embed at least 1 in. into studs. Install the panel at a comfortable height, so it will be easy to wire and access. Remove a concentric knockout in the panel and install

LOCATING A SUBPANEL

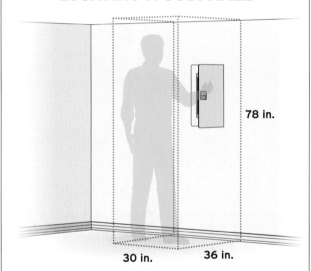

78 in.

30 in.   36 in.

Any panel—whether a main or a subpanel—must have sufficient clearance around it. NEC minimums require 78 in. of headroom, 36 in. free space in front of the panel, and 30 in. across the face of the panel. The panel should be installed at a comfortable height, meaning that no breaker handle may be higher than 72 in.

The area must also be dry and easily accessed. Do not install a subpanel near a bathroom or near flammable materials. If you install a subpanel in a common area between living spaces and an attached garage, wrap the panel enclosure with two layers of $\frac{5}{8}$ in. drywall on all sides. However, if you place the panel in a dry interior wall you can install a standard two-lug panel, whose 14$\frac{1}{2}$ in. width fits neatly between two studs spaced 16 in. on-center.

1 When possible, install a subpanel between studs spaced 16 in. on-center. Secure it in place using at least four screws.

4 Connect the neutral wire to the neutral bus lug. Plastic mounting brackets isolate—and insulate—neutral buses from the panel.

a connector appropriate to the cable or conduit that feeds the panel. Also install a protective cable bushing. Here, the feed is a #2 Romex cable. Pull 3 ft. of cable into the panel, tighten the connector, and strip the cable sheathing ❷.

In the photos, note the cross-brace between the studs, just below the panel. Code requires that you secure the cable within 12 in. of the panel, but 2/0 cable is stiff. Strapping cable to a stud would force it to enter the panel at a sharp angle, which could stress the knockout. It is far easier to strap the cable to a brace in the middle of the stud bay and run the cable straight into the panel.

Secure the ground wire to the main lug of the subpanel ground bus ❸. Cut the three insulated feed wires, allowing enough extra to loop them gently. Avoid sharp bends, which can damage wire. Strip 1 in. of insulation from the feed ends. Use an Allen wrench or torque wrench to connect the feed wires to their respective lugs ❹. Loop the hot feeds around the perimeter of the panel, distancing them from the neutral feed so there will be a open area through which to run smaller circuit wires ❺. Looping the wire generously also ensures that there will be enough extra cable in case you need to strip and reconnect feed ends later ❻.

ACCORDING TO CODE
The NEC requires a main breaker for any subpanel in a separate building.

2 Remove a knockout, install a cable connector, and then pull the feeder cable into the subpanel. Strap the cable to a cross-brace.

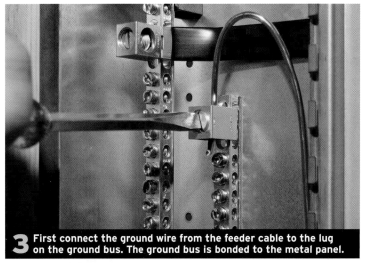

3 First connect the ground wire from the feeder cable to the lug on the ground bus. The ground bus is bonded to the metal panel.

5 Run the two hot feeds around the other side of the panel. Strip the hot feed ends and attach them to hot bus lugs.

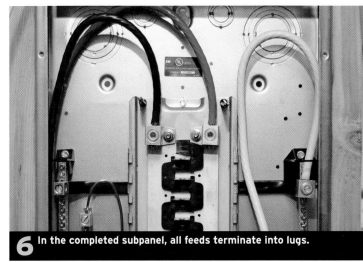

6 In the completed subpanel, all feeds terminate into lugs.

GLOSSARY

AC Armored cable.

ACCESSIBLE Not permanently concealed by the structure or finishes. Able to be accessed without damaging the building.

AMPACITY (AMPS) The amount of current a circuit or conductor can safely carry (conduct). Measured in amperes (amps).

AMPS The measure of the volume of electrons flowing through a system (current).

ARC FAULT CIRCUIT INTERRUPTER A circuit breaker designed to de-energize a circuit within a period of time if it senses arcing. An arc is a spark between conductors or connections.

BONDING JUMPER (MAIN) Conductor connecting the neutral conductor to the grounding electrode conductor in a Service Panel.

CABLE An assembly of several conductors. Most often refers to plastic- or metal-sheathed cable that contains several individual wires.

CIRCUIT (BRANCH) Circuit originating at a circuit breaker or fuse in an electrical panel and feeding utilization equipment (lights, switches, receptacles, appliances).

CIRCUIT BREAKER A device intended to de-energize a circuit if current exceeds specified parameters.

CONDULET A conduit body with a cover designed act as a pulling point in a conduit run, to change conduit direction, or as an intersecting point of multiple conduit runs.

CONDUCTOR Technically, anything that conducts electricity. Most often, it denotes a wire.

CURRENT The flow of electrons in a system. Current is measured in amperes (amps). There are two types of current: DC (direct current) and AC (alternating current). Typically AC is found in homes and buildings.

ENERGIZED Live The presence of voltage on a circuit or conductor.

FEEDER A large ampacity circuit supplying a Panel, Sub- Panel, or other high current piece of equipment.

FIXTURE A light fixture (table lamp, sconce, chandelier, recessed can etc.)

FUSE BOX A metal box designed to house fuses installed for circuit protection.

GROUND A connection to earth (the ground).

GROUND BUSS A piece of metal designed to connect multiple wires to a grounding electrode conductor.

GROUNDING CONDUCTOR A wire in an electrical system designed to bond metal parts of the electrical system to the earth (ground).

GROUNDING ELECTRODE CONDUCTOR A wire of gauge determined by the ampacity of a Service Panel that connects the service panel (and ground buss) to a grounding electrode (ground rod or equivalent).

GROUND FAULT A fault situation in which an energized conductor or piece of equipment comes in contact with grounded metal parts (ground).

GROUND FAULT CIRCUIT INTERRUPTER A special receptacle or circuit breaker intended to protect people by de-energizing a circuit within a specific period of time if current to ground exceeds specified parameters.

GROUND SCREW A green colored screw used to connect ground conductors to boxes and devices.

LISTED Equipment and materials included on a list published by an organization, acceptable to the authority having jurisdiction, that states that the equip-ment or materials meet specific design criteria or are suitable for the use intended. Most commonly UL (Underwriter's Laboratory) listed.

LOCATION, DAMP Protected from weather and not subject to saturation (under eves, canopies, etc.)

LOCATION, DRY Not subject to moisture (interior, protected from weather).

LOCATION, WET Underground; exposed to weather; subject to saturation.

MAIN CIRCUIT BREAKER A large ampacity circuit breaker that protects and acts as a means to de-energize a service panel or sub-panel.

MC Metal clad cable.

NEUTRAL The neutral or grounded conductor (not ground conductor) is the return path for current in an electrical system; designated by a white or light gray color. The neutral is bonded to ground at the Service Panel.

NEUTRAL BUS A piece of metal designed to connect multiple wires to the neutral conductor.

OHMS The measure of resistance to the flow of electrons (current).

OVERLOAD A situation in which the current flowing through a circuit or circuit conductor exceeds the safe operating ampacity of the conductor.

RECEPTACLE Device designed for the connection of an attachment plug.

ROMEX® Type NM cable, non-metallic sheathed.

SERVICE PANEL (MAIN PANEL) The primary circuit breaker or main breaker panel located immediately after the meter socket. It will contain a main breaker or disconnect means.

SERVICE Drop Conductors from the utility pole to the point of connection (weather head).

SERVICE-ENTRANCE CONDUCTORS (OVERHEAD) Conductors between the point of connection (meter lugs) and a point outside (weather head) connected to the service drop.

SERVICE-ENTRANCE CONDUCTORS (UNDERGROUND) The conductors between the point of connection (meter socket) and the utility service lateral.

SCREW SHELL The metal conductive interior body of a light socket or fuse holder with large threads (or ridges) that allow a bulb or fuse to be screwed in.

SPLICES The point at which two wires or conductors are joined or connected together. For branch circuits, splices are commonly made with pressure connectors (wire nuts).

SUB-PANEL An electrical panel that is supplied from a Service Panel and is installed either to provide additional circuit breakers or to distribute branch circuits at a distance from the Service Panel.

SWITCH A device indeed to open a circuit to control a light fixture, fan, or other piece of equipment or appliance.

VOLTAGE, NOMINAL (VOLTS) A value assigned to a circuit or system to designate its class. Modern single family residences in the U.S. typically have 120/240V systems.

VOLTAGE TO GROUND (VOLTS) The difference in potential (voltage) between a single energized conductor and ground (neutral or equipment ground).

WATERTIGHT Enclosures built to prevent the intrusion of moisture.

WATTAGE (WATTS) A measure of Power consumed.

WEATHERPROOF Equipment or enclosures built so that weather will not interfere with their operation.

INDEX